Handbook of Big Data Analytics and Forensics

Kim-Kwang Raymond Choo • Ali Dehghantanha
Editors

Handbook of Big Data Analytics and Forensics

 Springer

Editors
Kim-Kwang Raymond Choo 🆔
Department of Information Systems
and Cyber Security
The University of Texas at San Antonio
San Antonio, TX, USA

Ali Dehghantanha 🆔
Cyber Science Lab
School of Computer Science
University of Guelph
Guelph, ON, Canada

ISBN 978-3-030-74755-8   ISBN 978-3-030-74753-4   (eBook)
https://doi.org/10.1007/978-3-030-74753-4

This Springer imprint is published by the registered company Springer Nature Switzerland AG
The registered company address is: Gewerbestrasse 11, 6330 Cham, Switzerland

Acknowledgments

This book would not have been possible without the commitment of the contributing authors, who dedicated their time and efforts to research work and shared their findings in this book.

We are also extremely grateful to Springer and their staff for their support in this project. They have been most accommodating of our schedule and helping to keep us on track.

Contents

Big Data Analytics and Forensics: An Overview

Hossein Mohammadi Rouzbahani, Ali Dehghantanha ⓘ,
and Kim-Kwang Raymond Choo ⓘ

1 Introduction

As our society becomes smarter and more digitally connected, more data will be
generated, processed, disseminated, analyzed, and stored (e.g., on cloud computing
systems). Such big data can be structured and unstructured, and are generated by
different sources (e.g., Internet of Things (IoT) devices and other information and
communications technologies – ICT) with varying formats [1]. There have been
a number of definitions for big data, and one popular definition is the 5Vs model
where the 5Vs are Volume, Velocity, Variety, Veracity and Value [2]. Volume and
velocity refer to the size and formation speed of information respectively, while
variety refers to the diversity in data format and representation type. Veracity
refers to the accuracy and reliability of data, and finally, value attempts to quantify
usefulness of the data.

In a typical smart city setting, for example, IoT devices and other systems (e.g.,
edge/fog computing devices and servers) collect and process data before sending
them to cloud-based systems via high speed communication networks [3–6]. Hence,
it is important to ensure security for both data-at-rest and data-in-transit at the

H. M. Rouzbahani (✉)
Smart Cyber Physical Systems Lab, School of Engineering, University of Guelph,
Guelph, ON, Canada
e-mail: hmoham15@uoguelph.ca

A. Dehghantanha
Cyber Science Lab, School of Computer Science, University of Guelph, Guelph, ON, Canada
e-mail: adehghan@uoguelph.ca

K.-K. R. Choo
Department of Information Systems and Cyber Security, The University of Texas at San Antonio,
San Antonio, TX, USA
e-mail: raymond.choo@fulbrightmail.org

© Springer Nature Switzerland AG 2021
K.-K. R. Choo, A. Dehghantanha (eds.), *Handbook of Big Data Analytics
and Forensics*, https://doi.org/10.1007/978-3-030-74753-4_1

1

various devices and communication channels. Hence, we need solutions to help us perform big data analytics, security and forensic investigation to identify and extract relevant information, identify malicious activities and evidence of relevance, etc. [7].

Given the current trend in artificial intelligence, machine learning and deep learning, there have also been attempts to build on the advances in these areas to enhance security and forensic capabilities. For example, contemporary and emerging big data analytics approaches include generative-, discriminative- and hybrid-based methods. Recurrent Neural Network (RNN) and Long Short-Term Memory (LSTM) are two examples of supervised learning-based methods, which can facilitate the identification of movement pattern and human activity, as well as mobility prediction [8–12]. RNN is also useful in IoT applications with time dependent information, and processing series of data via internal memory. LSTM generally is known to achieve better performance with data of long time lag, and access to memory is protected by gates. Convolutional Neural Network (CNN) is a supervised learning technique with various IoT applications, including those that require large dataset for visual tasks (e.g., detecting patterns via feature extraction) [13–16].

The procedure of detecting, collecting, storing, analyzing and presenting of big data is also referred to as big data forensic. However, big data forensics is challenging, particularly if we also need to preserve user privacy. These challenges can be technical (use of strong encryption algorithms, the volume and veracity of data to be processed, etc.), legal (e.g., evidence and privacy legislations), and due to resources (or lack of), etc. [6, 17]. Several of these challenges will also be discussed in this book.

2 Book Outline

We will now describe the remaining 17 chapters.

Chapter "IoT Privacy, Security and Forensics Challenges: An Unmanned Aerial Vehicle (UAV) Case Study" [18] reviewed the existing literature on IoT security, privacy, and forensics. Next in Chapter "Detection of Enumeration Attacks in Cloud Environments Using Infrastructure Log Data" [19], the authors also explained how LSTM and CNN can be utilized to detect enumeration attacks in cloud-based environments. Chapter "Cyber Threat Attribution with Multi-view Heuristic Analysis" [20] presented a multi view heuristic analysis of malware by taking multiple characteristics of the malware files. Chapter "Security of Industrial Cyberspace: Fair Clustering with Linear Time Approximation" [21] presented a comparative summary of the performance for the different algorithms used to secure industrial cyberspace.

Then, Chapter "Adaptive Neural Trees for Attack Detection in Cyber Physical Systems" [22] demonstrated how decision tree and neural network can be combined to facilitate attack detection in Cyber Physical Systems (CPSs). In Chapter

"Evaluating Performance of Scalable Fair Clustering Machine Learning Techniques in Detecting Cyber Attacks in Industrial Control Systems" [23], a scalable fair clustering algorithm was utilized to build the Fair-let Decomposition (FD) model. Chapter "Fuzzy Bayesian Learning for Cyber Threat Hunting in Industrial Control Systems" [24] presented a machine learning algorithm which combines fuzzy logic with Bayesian inference to produce an optimized fuzzy model for identifying threats. In Chapter "Cyber-Attack Detection in Cyber-Physical Systems Using Supervised Machine Learning" [25], four different supervised learning methods – K-Nearest Neighbors (KNN), Support Vector Machine (SVM), Decision Tree (DT), and Random Forest (RF) – were employed to build models to detect cyber-attack activities on a water treatment plant.

Evaluation of scalable fair clustering machine learning methods for threat hunting in CPSs were presented in Chapter "Evaluation of Scalable Fair Clustering Machine Learning Methods for Threat Hunting in Cyber-Physical Systems" [26], while two supervised and unsupervised machine learning classifiers for Mac OS malware detection were the focus in Chapter "Evaluation of Supervised and Unsupervised Machine Learning Classifiers for Mac OS Malware Detection" [27]. The effectiveness of different machine learning methods (e.g., Random Forest, KNN, DT, Naïve Bayes, and SVM) on IoT malware opcodes and Mac OS X malware detection was presented in Chapter "Evaluation of Machine Learning Algorithms on Internet of Things (IoT) Malware Opcodes" [28] and Chapter "Mac OS X Malware Detection with Supervised Machine Learning Algorithms" [29], respectively. Chapter "Machine Learning for OSX Malware Detection" [30] presented a summary of the performance for different learning-based approaches in OSX malware detection.

Chapter "Hybrid Analysis on Credit Card Fraud Detection Using Machine Learning Techniques" [31] compared the performance of different deep learning, supervised, unsupervised and hybrid learning techniques in credit card fraud detection. Chapter "Mapping CKC Model Through NLP Modelling for APT Groups Reports" [32] introduced an automated way of processing Advanced Persistent Threats (APT) reports to identify and map the different Cyber Kill Chain (CKC) stages employed in the attack. Chapter "Ransomware Threat Detection: A Deep Learning Approach" [33] utilized the performance of five different machine learning techniques (i.e., KNN, CNN, DT, logistic regression and random forest) in ransomware threat detection. Chapter "Scalable Fair Clustering Algorithm for Internet of Things Malware Classification" [34] studied the effect of scalable clustering algorithm on accuracy, by experimenting with a IoT malware opcodes dataset.

References

1. H.M. Rouzbahani, H. Karimipour, A. Rahimnejad, A. Dehghantanha, G. Srivastava, Anomaly detection in cyber-physical systems using machine learning, in *Handbook of Big Data Privacy*, ed. by K.-K. R. Choo, A. Dehghantanha, (Springer, Cham, 2020), pp. 219–235

2. H. Mohammadi Rouzbahani, H. Karimipour, G. Srivastava, Big data application for security of renewable energy resources, in *Handbook of Big Data Privacy*, ed. by K.-K. R. Choo, A. Dehghantanha, (Springer, Cham, 2020), pp. 237–254
3. M. Mohammadi, A. Al-Fuqaha, S. Sorour, M. Guizani, Deep learning for IoT big data and streaming analytics: A survey. IEEE Commun. Surv. Tutor. **20**(4), 2923–2960 (2018). https://doi.org/10.1109/COMST.2018.2844341
4. M. Conti, T. Dargahi, A. Dehghantanha, *Cyber Threat Intelligence: Challenges and Opportunities* (Springer, Cham, 2018)
5. H. HaddadPajouh, R. Khayami, A. Dehghantanha, K.-K.R. Choo, R.M. Parizi, AI4SAFE-IoT: An AI-powered secure architecture for edge layer of Internet of things. Neural Comput. & Applic., 1–15 (2020, February). https://doi.org/10.1007/s00521-020-04772-3
6. A. Azmoodeh, A. Dehghantanha, K.-K.R. Choo, Big data and internet of things security and forensics: Challenges and opportunities, in *Handbook of Big Data and IoT Security*, (Springer, Cham, 2019), pp. 1–4
7. S. Srinivasan, Security and privacy in the computer forensics context, in *2006 International Conference on Communication Technology*, (Guilin, China, 2006, November), pp. 1–3. https://doi.org/10.1109/ICCT.2006.341936
8. H.M. Rouzbahani, Z. Faraji, M. Amiri-Zarandi, H. Karimipour, AI-enabled security monitoring in smart cyber physical grids, in *Security of Cyber-Physical Systems*, ed. by H. Karimipour, P. Srikantha, H. Farag, J. Wei-Kocsis, (Springer, Cham, 2020), pp. 145–167
9. M. Saharkhizan, A. Azmoodeh, A. Dehghantanha, K.-K.R. Choo, R.M. Parizi, An ensemble of deep recurrent neural networks for detecting IoT cyber attacks using network traffic. IEEE Internet Things J. **7**(9), 8852–8859 (2020, September). https://doi.org/10.1109/JIOT.2020.2996425
10. H.H. Pajouh, A. Dehghantanha, R. Khayami, K.-K.R. Choo, A deep recurrent neural network based approach for Internet of Things malware threat hunting. Futur. Gener. Comput. Syst. **85**, 88–96 (2018). https://doi.org/10.1016/j.future.2018.03.007
11. S. Homayoun et al., DRTHIS: Deep ransomware threat hunting and intelligence system at the fog layer. Futur. Gener. Comput. Syst. **90**, 94–104 (2019). https://doi.org/10.1016/j.future.2018.07.045
12. K.C.A.N. Jahromi, S. Hashemi, A. Dehghantanha, R. Parizi, An enhanced stacked LSTM method with no random initialization for malware threat hunting in safety and time-critical systems. IEEE Trans. Emerg. Top. Comput. Intell. **4**(5), 630–640 (2020)
13. H.M. Rouzbahani, H. Karimipour, L. Lei, An ensemble deep convolutional neural network model for electricity theft detection in smart grids. Presented at the Future Technologies Conference (FTC), 2020
14. A. Azmoodeh, A. Dehghantanha, R.M. Parizi, S. Hashemi, B. Gharabaghi, G. Srivastava, Active spectral botnet detection based on eigenvalue weighting, in *Handbook of Big Data Privacy*, ed. by K.-K. R. Choo, A. Dehghantanha, (Springer, Cham, 2020), pp. 385–397
15. A. Azmoodeh, A. Dehghantanha, Big data and privacy: Challenges and opportunities, in *Handbook of Big Data Privacy*, ed. by K.-K. R. Choo, A. Dehghantanha, (Springer, Cham, 2020), pp. 1–5
16. M. Conti, A. Dehghantanha, K. Franke, S. Watson, Internet of Things security and forensics: Challenges and opportunities. Futur. Gener. Comput. Syst. **78**, 544–546 (2018). https://doi.org/10.1016/j.future.2017.07.060
17. G.S. Chhabra, V.P. Singh, M. Singh, Cyber forensics framework for big data analytics in IoT environment using machine learning. Multimed. Tools Appl. **79**(23–24), 15881–15900 (2020, June). https://doi.org/10.1007/s11042-018-6338-1
18. I. Lazo, A. Pardo, E. Patch, A. Dehghantanha, K.-K.R. Choo, IoT privacy, security and forensics challenges: An unmanned aerial vehicle (UAV) case study, in *Handbook of Big Data Analytics and Forensics*, ed. by A. Dehghantanha, K. K. R. Choo, (Springer, Cham, this volume)

19. S. Gharghashe Eisaloo, T. Steinbach, Detection of enumeration attacks in cloud environments using infrastructure log data, in *Handbook of Big Data Analytics and Forensics*, ed. by A. Dehghantanha, K. K. R. Choo, (Springer, Cham, this volume)
20. D.K. Sahoo, Cyber threat attribution with multi-view heuristic analysis, in *Handbook of Big Data Analytics and Forensics*, ed. by A. Dehghantanha, K. K. R. Choo, (Springer, Cham, this volume)
21. N. Chikhalia, Y. Dhawan, Security of industrial cyberspace: Fair clustering with linear time approximation, in *Handbook of Big Data Analytics and Forensics*, ed. by A. Dehghantanha, K. K. R. Choo, (Springer, Cham, this volume)
22. C. Chen, K. Wulff, Adaptive neural trees for attack detection in cyber physical systems, in *Handbook of Big Data Analytics and Forensics*, ed. by A. Dehghantanha, K. K. R. Choo, (Springer, Cham, this volume)
23. A. Handa, P. Semwal, Evaluating performance of scalable fair clustering machine learning techniques in detecting cyber attacks in industrial control systems, in *Handbook of Big Data Analytics and Forensics*, ed. by A. Dehghantanha, K. K. R. Choo, (Springer, Cham, this volume)
24. K. Marsha, E.G. Samira, Fuzzy Bayesian learning for cyber threat hunting in industrial control systems, in *Handbook of Big Data Analytics and Forensics*, ed. by A. Dehghantanha, K. K. R. Choo, (Springer, Cham, this volume)
25. P. Semwal, A. Handa, Cyber-attack detection in cyber-physical systems using supervised machine learning, in *Handbook of Big Data Analytics and Forensics*, ed. by A. Dehghantanha, K. K. R. Choo, (Springer, Cham, this volume)
26. D. Sahoo, A. Upadhyay, Evaluation of scalable fair clustering machine learning methods for threat hunting in cyber-physical systems, in *Handbook of Big Data Analytics and Forensics*, ed. by A. Dehghantanha, K. K. R. Choo, (Springer, Cham, this volume)
27. D. Sahoo, Y. Dhawan, Evaluation of supervised and unsupervised machine learning classifiers for Mac OS malware detection, in *Handbook of Big Data Analytics and Forensics*, ed. by A. Dehghantanha, K. K. R. Choo, (Springer, Cham, this volume)
28. A. Anidu, Z. Obuzor, Evaluation of machine learning algorithms on Internet of Things (IoT) malware opcodes, in *Handbook of Big Data Analytics and Forensics*, ed. by A. Dehghantanha, K. K. R. Choo, (Springer, Cham, this volume)
29. S.E. Gharghasheh, S. Hadayeghparast, Mac OS X malware detection with supervised machine learning algorithms, in *Handbook of Big Data Analytics and Forensics*, ed. by A. Dehghantanha, K. K. R. Choo, (Springer, Cham, this volume)
30. C. Chen, K. Wulff, Machine learning for OSX malware detection, in *Handbook of Big Data Analytics and Forensics*, ed. by A. Dehghantanha, K. K. R. Choo, (Springer, Cham, this volume)
31. A. Handa, Y. Dhawan, P. Semwal, Hybrid analysis on credit card fraud detection using machine learning techniques, in *Handbook of Big Data Analytics and Forensics*, ed. by A. Dehghantanha, K. K. R. Choo, (Springer, Cham, this volume)
32. A. Upadhyay, S.E. Gharghasheh, S. Nakhodchi, Mapping CKC model through NLP modelling for APT groups reports, in *Handbook of Big Data Analytics and Forensics*, ed. by A. Dehghantanha, K. K. R. Choo, (Springer, Cham, this volume)
33. K. Marsh, A. Handa, H. Haddadpajouh, Ransomware threat detection: A deep learning approach, in *Handbook of Big Data Analytics and Forensics*, ed. by A. Dehghantanha, K. K. R. Choo, (Springer, Cham, this volume)
34. Z. Obuzor, A. Anidu, Scalable fair clustering algorithm for Internet of Things malware classification, in *Handbook of Big Data Analytics and Forensics*, ed. by A. Dehghantanha, K. K. R. Choo, (Springer, Cham, this volume)

IoT Privacy, Security and Forensics Challenges: An Unmanned Aerial Vehicle (UAV) Case Study

Isis Diaz Linares, Angelife Pardo, Eric Patch, Ali Dehghantanha ⓘ, and Kim-Kwang Raymond Choo ⓘ

1 Introduction

Internet of Things or "IoT" can be broadly defined as the convergence of the Internet and smart objects that communicate and interact with each other, gather information, and analyze it to complete a task or learn from a process [1, 2]. The rapid development of IoT devices creates an enormous amount of data transmission over unprotected systems and introduces numerous vulnerabilities impacting forensics, security and privacy. Many of these IoT devices lack firewalls, antivirus software and intrusion detection systems in the programming of the product and pose multiple risks to users [3–5].

In a study conducted by Hewlett Packard, 84% of worldwide organizations who have adopted IoT devices have already experienced a security breach [6, 7]. Breaches can lead to data loss, data corruption, denial of access and/or complete device takeover [8–11]. Samsung's Open Economy report identifies the criticality of securing every connected IoT device, approximately 7.3 billion, by 2020 [12–15]. In the same manner, Casey encourages digital investigators to become more familiar with IoT technology and understand the kinds of evidence they contain involving criminal activity [2, 16–22].

One candidate for demonstrating the various facets of IoT's underlying technologies and capabilities is drone (also referred to as unmanned aerial vehicle (UAV)); drones are fundamentally an amalgam of every IoT technology, along

I. Diaz Linares · A. Pardo (✉) · E. Patch · K.-K. R. Choo
Department of Information Systems and Cyber Security, The University of Texas at San Antonio, San Antonio, TX, USA
e-mail: raymond.choo@fulbrightmail.org

A. Dehghantanha
Cyber Science Lab, School of Computer Science, University of Guelph, Guelph, ON, Canada
e-mail: adehghan@uoguelph.ca

© Springer Nature Switzerland AG 2021
K.-K. R. Choo, A. Dehghantanha (eds.), *Handbook of Big Data Analytics and Forensics*, https://doi.org/10.1007/978-3-030-74753-4_2

with Cyber Physical Engineering Systems (CPES). They incorporate a network of sensors, cameras, storage, CPUs and actuators with various network connections including Cellular, GPS, Wi-Fi, Bluetooth, and radio frequency that enable them to share data and control information in numerous ways that a set of vulnerabilities and security concerns are associated to each component [23–33]. Drones [34] are quickly becoming ubiquitous in everyday life; each year they are smaller, cheaper, and easier to fly, making them more appealing to use. They have also proven to be a viable solution to many tasks and have become an essential tool in areas such as gathering data for disaster management, reconnaissance of remote areas for inspections of things such as pipelines, aerial photography/cinematography, fast shipping delivery for companies such as Amazon, geographic mapping, crop monitoring, building/structure safety inspection, military weapons, law enforcement, crime such as smuggling, spying, and even assassinations.

The purpose of this chapter is to present a comparative study of the relationship between the challenges connected to security, forensics, and privacy with IoT devices. Existing literature available about IoT devices reveal considerable overlaps with the challenges in forensics, security and privacy, but a study of their interrelatedness has not been explored.

2   Proposed Research Methodology

The work presented in this chapter has been organized in a systematic review to identify current research efforts about the challenges of security, privacy, and forensics with IoT devices. The research question that guided the review was: "What is the relationship between the challenges of security, forensics, and privacy in the IoT environment?"

Our research method is limited to literature documenting challenges in 'IoT' and only included literature published after 2000 in order to capture the most recent trends in IoT. Relevant literature was identified through various major databases available through the University: ACM Digital Library, IEEE Xplore/IET Electronic Library Online, Safari Books Online, Research Direct and Semantics Scholar. The search terms intended to identify all literature that covered security, privacy, and forensics challenges were: "internet of things or IoT", "challenges in security of internet of things", "challenges in privacy of internet of things", "challenges in forensics of internet of things", "drones", "Unmanned Aerial Vehicle", "UAV", "Unmanned Aerial System", "UAS", "Cyber Physical Engineering Systems" and "CPES". The identified literature was then analyzed to determine relevance in the research project. The title, abstract and content of the chosen papers were analyzed to determine if they were relevant or not. Irrelevant papers were removed from the review list in this phase. To arrive at our final selections, the publications were analyzed to answer the research question of this paper. Software tools such as Google Docs, Microsoft Excel and Microsoft Word were used to compile and analyze this collection of literature. The studies were organized according to the challenges in security, privacy and forensics in a literature review. After validating consistency and reliability, the relevant articles and research manuscripts were chosen in order to analyze the challenges posed to the traditional models of forensics, security and privacy with IoT devices.

3 Results and Discussion

In this section we present a literature review divided by the characteristics of IoT that create challenges to Forensics, Security and/or Privacy as identified by existing literature. In addition, despite a source's focus on only 1 of the 3 models (Forensics, Security or Privacy), we identify whether each characteristic also affects the other 2 models. The literature review is summarized in Table 1. In Sect. 3.5 and Fig. 1 we then demonstrate how these characteristics provided the basis for the consolidated list of 15 common challenges related to forensics, security and privacy of IoT devices.

3.1 Challenges That IoT Devices Introduce to the Traditional Forensics Model

The rise of crime, encouraged by the vulnerabilities of IoT devices, calls on effective digital forensic processes to gather forensically-sound evidence for a case under investigation. Forensics analysis and methodologies are very new: one of the first methodologies was proposed in 2013 and the first proposed definition of IoT forensics was made as recently as 2015 [25]. According to McKemmish, "digital forensics is the process of identifying, preserving, analyzing, and presenting digital evidence in a manner that is legally acceptable" [12, 52]. It has four key elements: Identification, Preservation, Analysis, and Presentation [12].

3.1.1 Challenges in Identification

The identification phase involves the tasks of identifying and managing possible sources of evidence. Unlike traditional devices that contain storage media (i.e. computers, servers, etc. [53, 54]), data collection could occur across a diverse set of IoT devices. This can also create challenges with interoperability, compatibility and integration, which could negatively affect data extraction, interpretation and verification [29, 52]. This added complexity makes the mechanisms to collect and analyze data of IoT devices even more critical [55] in a forensics investigation.

According to Oriwoh et al. and Baig et al., evidence data such as usage history may be stored in multiple locations causing an expansion of evidence sources

Table 1 Literature review summary

Writers	Article title	Characteristic of IoT that creates challenges in Forensics	Security	Privacy
Watson & Dehghantanha [11]	Digital forensics: the missing piece of the Internet of Things promise	Data storage is not accessible (Cloud w/o 3rd party agreement or overseas)		
Conti et al. [28]	Internet of Things security and forensics: Challenges and opportunities	Real-time and autonomous interactions between different nodes make preservation of evidence difficult		
		No metadata available for analysis such as modified, accessed, created time		
		Absence of proper authentication makes it difficult to identify responsible parties	X	
		No logging or monitoring system	X	
Casey, Eoghan [13]	The value of forensic preparedness and digital-identification expertise in smart society	Unreliability of big data due to errors or omission		
Baig et al. [35]	Future challenges for smart cities: Cyber-security and digital forensics	Variety in network and application architecture	X	
		Data exists in multiple locations (encrypted, stored locally or in the Cloud)		
		Difficult to acquire device/evidence		
		Data storage is not accessible (Cloud w/o 3rd party agreement or overseas) – must go through multiple jurisdictions		X

Oriwoh et al. [27]	Internet of Things Forensics: Challenges and approaches	Expansion of evidence sources		
		Expansion of multiple devices to devise conclusive facts		
		Inclusion of evidence not limited to standard file formats but dependent on vendor's data types		
		Large quantity of data to analyze		
		Undefined boundaries between device data and ownership	X	
		Data may not be readable w/existing tools		
Casey, Eoghan [13]	Smart Home Forensics	Limited access to metadata associated with digital evidence		X
		Usage history stored in multiple locations (device, cloud)		
		IFTTT recipes can make persons of interest highly trackable	X	X
Do et al. [36]	FAU Open Research Challenge: Digital Forensics Forensic Report	Passwords on the Raspbian OS can stay as the default "raspberry"	X	X
Armir Bujari & Furini [23]	Standards, Security and Business Models: Key Challenges for the IoT scenario	Interoperability and compatibility/integration affects data extraction (due to diversity of objects)		
		Data leakage and data manipulation	X	
		Personal information used to trade for services (i.e. mail account) can be exploited	X	X
Oriwoh, Edewede [37]	Internet of Things – the argument for Smart Forensics	Locating and gaining access to widely dispersed objects of forensics interest may be a challenge (Oriwoh 2013 reference)		
		Difficulty to identify sources of breaches due to IoT's autonomous systems and adaptability (data can be changed without human input)	X	X
		Variety between evidence sources (timestamp formats, packet headers, etc.) may cause confusion until they are unified to achieve useful correlation		
		Multiple access points (physical and virtual realm) for changing data may cause evidence tampering	X	X

(continued)

Table 1 (continued)

Writers	Article title	Characteristic of IoT that creates challenges in Forensics	Security	Privacy
		Accessing device and also acquiring evidence without destroying device because processors and storage are so small		
		Dependency increases on central home control servers, external service providers, network logs, and other user devices – which need to be determined in planning stage		
		Device may be located in different jurisdictions with varying laws and policies		X
		Continually adjusts and changes form (self-destruction)		
		Absence of industry-wide standards for data protocols, platforms, connectors, etc.		
Khan et al. [24]	A Comprehensive Review on Adaptability of Network Forensics Frameworks for Mobile Cloud Computing	Network forensics framework (NFF) is less reliable when there are too many resources for computation due to irrelevant data and time-consuming analysis		
Perumal et al. [26]	Internet of Things(IoT) digital forensic investigation model: Top-down forensic approach methodology	Size of the objects of forensic interest, relevancy, blurry network boundaries and edgeless networks		X
Awasthi et al. [28]	Welcome pwn: Almond smart home hub forensics	Challenges include extracting, accessing, interpreting and verifying the data		
		No removable media in IoT device or smart hub		
cline3-5		Data is lost when device is powered off		
Kalaimannan, Ezhil [29]	Smart Device Forensics – Acquisition, Analysis and Interpretation of Digital Evidences	Forensics challenges include nature of file systems, logical memory, data structures, and third party applications that can be installed or accessed by the device		
Do et al. [25]	Cyber-physical systems information gathering: A smart home case study	Difficulty in accessing evidential data: 3rd party vendors sometimes control access to the data and may be unwilling to comply		X

Reference	Title	Description		
Babun et al. [30]	IoTDots: A Digital Forensics Framework for Smart Environments	Researchers have shown that a variety of Smart Home devices are vulnerable to different attacks: the trust model of smart locks could be used against the smart lock to allow an intruder to bypass a status of revoked access, malicious data could be transferred via smart lights by manipulating the intensity of the light in a way that a user would be unaware of the manipulation	X	
		Forensics analysis and methodologies are very new: one of the first methodologies was proposed in 2013 and the first proposed definition of IoT forensics was made as recently as 2015		
		Manufacturers are neglecting to provide countermeasures in most smart home devices to protect against LAN-based attacks (ex: there is no client authentication)	X	
		Even passive attackers can gain access to a significant amount of user data from smart home devices despite the devices' limited computing power (i.e. usage data, location data)	X	X
Risteska Stojkoska & Trivodaliev [31]	A review of Internet of Things for smart home: Challenges and solutions	No common standard due to variety of network protocols used IoT devices have limited power, computing		
		Resources and storage capacity		
		Data collection occurs across a diverse set of devices so mechanism to collect and analyze data is important		
		Challenges include data processing issues (energy, memory capacity and processing capabilities), networking and interoperability		
		Big data management in the cloud can affect data integrity	X	X

(continued)

Table 1 (continued)

Writers	Article title	Characteristic of IoT that creates challenges in Forensics	Security	Privacy
Zia et al. [32]	Application-Specific Digital Forensics Investigative Model in Internet of Things (IoT)	IoT is vulnerable to common attacks of wireless networks	X	
		Devices are heterogenous and data source may be on the cloud instead of on device		
		Lack of unified standards		
		Identity security, data security, behavior security	X	
		Hardware limitations include computational, energy, memory, and physical.		
		Software limitations include embedded software, dynamic security patch, mobility, scalability, multiplicity, and dynamic nature of networks.	X	X
		Forensics may be negatively impacted by physical tampering, wireless or RF interference, rogue device inserted in same network, or malicious code injection	X	X
		Lack of encryption in communication and downloading software updates	X	X

Writers	Article title	Characteristic of IoT that creates challenges in security	Forensics	Privacy
Plachkinova et al. [33]	Emerging Trends in Smart Home Security, Privacy, and Digital Forensics	Lack of adequate access right administration	X	
		Remote Access Tools		X
		Vulnerable ecosystems (i.e. Android; leak sensitive data)		X
		Lack of security keys	X	

Zarpelao et al. [1]	A survey of intrusion detection in Internet of Things	IoT devices typically rely on resource-constrained nodes restricting the implementation of IDS	X	X
		Security and Privacy countermeasures are difficult to implement because components of IoT devices have limited computing power, the intent of the connections between devices is to share data not to avoid sharing it, and the sheer quantity of connections	X	X
		The fast productization of IoT (speed at which products are being created stemming from the idea of IoT)	X	X
		IoT technology vendors are responsible for releasing patches to address vulnerabilities		
		IoT security threats fall into the following categories: copying data, removing data and substituting with purposely corrupt or malicious data, replacing firmware, removing security countermeasures, eavesdropping, man-in-the-middle attack, routing attack, and DoS	X	X
		Signature-based intrusion detection is unsuitable to IoT (because of limited computing power)		
Plachkinova et al. [33]	Emerging Trends in Smart Home Security, Privacy, and Digital Forensics	System administrators typically lack expertise		
		Visitors to the home may have unsecured access to smart devices and may impact usage data	X	X
		Devices are not used by the same people in the same manner from one smart home to another (i.e. some homes have persons who are more vulnerable to social engineering conducted via smart home devices)	X	
		There is no government authority holding manufacturers accountable for providing a minimum level of security protection in smart home devices	X	X
		There are no requirements for how usage data is stored	X	X
		Manufacturers are not accountable to any specific data retention policies	X	X

(continued)

Table 1 (continued)

Writers	Article title	Characteristic of IoT that creates challenges in Forensics	Security	Privacy
Schiefer, Michael [38]	Smart Home Definition and Security Threats	Consumer education/knowledge: Smart Home devices are often miscategorized because there is no universal definition of Smart Home		
		Use of factory-default passwords	X	X
		None or insufficient encryption for communication	X	X
		A smart home is only as safe as its weakest device	X	X
Sicari et al. [39]	Security, privacy and trust in Internet of Things: The road ahead	Different standards among IoT devices		
		Scalability of security measures		
		Limited computing power of IoT devices	X	
Alaba et al. [40]	Internet of Things security: A survey	Categorizes security challenges as: privacy, authorization, verification, access control, system configuration, information storage and management		X
		IoT devices are set up on Low Power and Lossy Networks (LLN), which are constrained by limited processing power	X	
		Limited computing power of IoT devices	X	
		Vulnerability to certain types attacks: man-in-the-middle, counterfeit		
Ashraf & Habaebi [41]	Autonomic schemes for threat mitigation in Internet of Things	IoT devices are so diverse that eventually it will be so complicated to maintain security for each type of device that devices need to be able to maintain security for themselves	X	X
		IoT relies on wireless communication which is significantly more vulnerable than wired communication	X	

Author	Title	Challenge		
		Limited computing power of IoT devices	X	
		IoT devices are particularly vulnerable to DoS attacks		
		No solution has been found to the vulnerability to man-in-the-middle attacks		
		Easier to design security measures for specific security issues, not for broader security		
		IoT devices often have a central processing router, which can be a single point of failure for the network	X	
Zhang et al. [42]	IoT Security: Ongoing Challenges and Research Opportunities	Diversity of IoT devices	X	
		Scalability of security measures		
		Manufacturer lack of accountability		
		Impossible to employ cryptographically pre-shared keys		
		No shared standards for Authentication and Authorization	X	
		Object Identification: IoT devices use DNS which is susceptible to man-in-the-middle attacks		
Sha et al. [43]	On security challenges and open issues in Internet of Things	IoT devices function off of limited resources such as limited computing power, energy supply and memory	X	
		There are IoT devices on the market that are low quality but can put users at significant risk based on the increased security vulnerabilities	X	
		Users care more about convenience than security		X
		Variety of IoT devices	X	

(continued)

Table 1 (continued)

Writers	Article title	Characteristic of IoT that creates challenges in Forensics	Security	Privacy
Heartfield et al. [44]	A taxonomy of cyber-physical threats and impact in the smart home	Heterogeneous technologies	X	
		Lack of user knowledge		X
		Limited computing power of IoT devices that limits authentication		
		Wireless communication	X	
		Manufacturers and users prefer lower costs devices than security		
Komminos et al. [45]	Survey in Smart Grid and Smart Home Security: Issues, Challenges and Countermeasures	User authentication	X	X
		Vulnerability to DoS attacks	X	
		Regulation is not standardized	X	
		Lack of authorities to hold manufacturers accountable for upholding security standards in IoT devices	X	
Lin & Bergmann [46]	IoT Privacy and Security Challenges for Smart Home Environments	Main challenges are confidentiality, authentication and access	X	X
		Lack of user knowledge or professional resources	X	X
		Vulnerability of wireless communication	X	
		Heterogeneous technologies	X	
		Lack of consistent software patches for security	X	
		Limited standards among manufacturers for security measures	X	
		Lack of professional services to focus on Smart Home Security		X

Reference	Title	Challenge		
Geneiatakis et al. [47]	Security and privacy issues for an IoT based smart home	Restricted computing power of IoT devices		X
		Vulnerabilities caused specifically because of the IoT architecture	X	X
		Heterogeneous technologies		X
		Lean operating systems that require fewer resources to compromise	X	X
		Users can purchase devices with malware pre-installed by malicious vendors		X
		Apps associated with smart home devices are a weak point in security	X	X
		Overreliance on the router's firewall in the home network		
Notray et al. [48]	An experimental study of security and privacy risks with emerging household appliances	Security measures taken by manufacturers and users are inconsistent	X	X
		Lack of user knowledge	X	X
		Scalability of security measures		X
		Limited computing power		X
		Power consumption restraints		X
Whittaker, Zach [49]	After massive cyberattack, shoddy smart device security comes back to haunt	Users don't care enough about security	X	X
		Manufacturers' use of factory-default passwords	X	X
		Attribution		X
		Users keeping factory-default passwords	X	X

Table 1 (continued)

Writers	Article title	Characteristic of IoT that creates challenges in Forensics	Security	Privacy
Ronen & Shamir [50]	Extended Functionality Attacks on IoT Devices: The Case of Smart Lights	Intended functionality of IoT devices causes vulnerabilities that back-fire		
		Some devices require a setup process where unencrypted passwords are communicated via Wi-Fi	X	X
		Lack of standard security measures used by manufacturers and designers	X	X
		IoT devices are sometimes unnecessarily integrated, creating weak points where these are not necessary for functionality	X	X

Writers	Article title	Characteristic of IoT that creates challenges in privacy	Security	Forensics
Plachkinova et al. [33]	Emerging Trends in Smart Home Security, Privacy, and Digital Forensics	Manufacturers often utilize usage data to either sell it to 3rd parties or to customize advertising to the user		
Lopez et al. [51]	Evolving privacy: From sensors to the Internet of Things	Users no longer need to take actions like searching for something online to have their privacy be at risk because smart devices are sensing their behaviors		X
		Device ownership becomes unclear as IoT scales		
		Loss of privacy cannot be undone		
		Lack of legislation		

with multiple access points for changing data [25, 29]. When data is stored in various locations, risk of evidence tampering through data leakage and manipulation become highly probable [52]. Locating and gaining access to a variety of objects of forensics interest could pose multiple challenges for user privacy and create conflict with jurisdiction. This heightened complexity makes it difficult to devise conclusive facts in a forensics investigation [25].

Watson and Dehghantanha also point out that data storage may not be located on the device but on a third-party cloud service and possibly even overseas [11, 56]. This makes it challenging to access evidential data since third-party vendors control access and may not be willing to give access to forensics investigators [25]. If the data or device is located overseas, varying laws and policies in different jurisdictions could also make the acquisition of evidence problematic [27, 29]. This raises other issues of undefined boundaries between device data and ownership [25].

3.1.2 Challenges in Analysis

The analysis phase is focused on examining and analyzing the data collected for evidence through the use of proper forensic tools. According to Zia et al., there

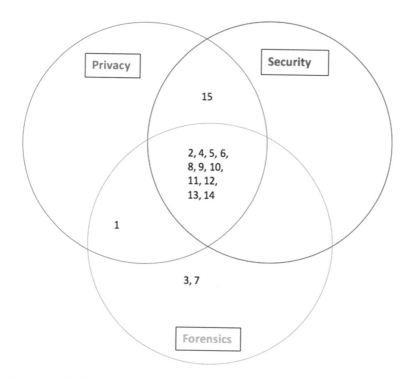

Fig. 1 How the challenges in forensics, security and privacy interrelate

Challenge #	Challenge Description
1	Data storage not accessible
2	Autonomous changes to data
3	Lack of metadata
4	Lack of secure authentication
5	Unreliable big data
6	Data in many locations
7	Difficult to acquire device
8	Diversity of objects
9	Various jurisdictions may result in legal conflict
10	No standards
11	Minimal features
12	Added security vulnerability
13	Difficult to scale security measures
14	Lack of user expertise
15	Other

Fig. 1 (continued)

is currently a lack of unified standards for IoT devices due to the variety of network protocols used [29, 31]. The absence of unification with data protocols, platforms and connectors create inefficiency and confusion throughout various evidence sources [27]. In addition, Oriwoh states that the inclusion of evidence is dependent on the vendor's data types which may not be standard file formats [25]. This results in a variety of timestamp formats, packet headers, and other key data types used for forensic evidence. This poses significant challenges to data analysis since non-standard data types may not be readable with existing forensics tools [25].

According to Oriwoh, IoT devices have limited power, computing resources and storage capacity [55], which cause limitations in the device's memory capacity, processing capabilities and interoperability [30, 56, 57]. Often times there is no removable media available [29] nor a logging or monitoring system [28]. Due to the limited storage and computing resources of IoT devices, challenges arise with accessing metadata associated with digital evidence [13]. There may not be metadata available such as dates and times for file modification, access, or creation [28]. This then leads the forensics investigator to depend on central home control servers, external service providers, network logs, and other devices owned by the user [27]. Do et al. asserts that an intruder could take advantage of the vulnerabilities that arise from the lack of secure authentication [25]. This absence of proper authentication makes it difficult to identify responsible parties according to Conti et al. [58].

Stojkoska et al. also claims that big data management in the cloud can affect data integrity [57] through errors and omission, which makes the big data unreliable and challenging to analyze [13, 25]. Khan et al. further illustrates that the network forensics framework (NFF) is less reliable when there are too many resources for computation due to irrelevant data and time-consuming analysis [59].

3.1.3 Challenges in Presentation

The presentation phase presents the evidence to courts through expert testimony on the analysis of the evidence. The issues identified in the identification, preservation and analysis phases may result in an incomplete or incorrect forensic investigation and thus directly impact the presentation phase. Until a more effective process to investigate IoT-related incidents is developed, criminals will benefit from the lack of reliable and accurate evidence.

3.2 Challenges That IoT Devices Introduce to the Traditional Security Model

Classically, the goal of Information Security has been to ensure Confidentiality, Integrity and Availability of information, making up the popular "CIA Triad" [25].

3.2.1 Challenges in Confidentiality

According to Bosworth et al. there are two ways of losing confidentiality: through disclosure and/or observation. Disclosure can involve the user intentionally sharing information with another user or unintentionally leaving the information unattended and visible, such as walking away from a computer monitor that is still turned on. Observation, on the other hand, involves an unauthorized person taking steps to access the information. Observation does not, by definition, have to be malicious. For example, if a user accesses his own bank account information on a public computer and turns off the screen to step away momentarily, another user may approach the computer and turn on the monitor without intending to compromise the confidentiality of the bank account information but still observe the information [25].

IoT devices face challenges stemming from the minimal features each device can support. These devices possess limited computing power, which then limits the authentication and encryption protocols that can be supported [39]. According to Zhang et al., it is impractical and many times impossible to use cryptographically pre-shared keys [42, 60]. According to Zarpelao et al., limited computing power also makes signature-based intrusion detection unsuitable for IoT devices [1].

IoT devices rely more heavily on wireless communication (as opposed to wired communication) [25]. This causes a significant increase in their vulnerability to certain types of attacks, such as man-in-the-middle attacks [1, 40, 42].

Zarpelao et al. discuss that the fast productization, or fast speed at which IoT devices are being developed and marketed, makes keeping up with security challenges a daunting task 1. This challenge is magnified by users' lack of expertise and even user apathy toward prioritizing security. Lack of expertise can lead to behaviors such as leaving manufacturer default passwords in effect, susceptibility to social engineering conducted via IoT devices, and overreliance on the router's firewall in the home network [33, 38, 48, 50]. Users often value the convenience of the features each IoT device offers higher than security [43].

3.2.2 Challenges in Integrity

The integrity of data is compromised when the data is no longer whole [25]. For example, if a user took a video file from a little-known producer, modified the file to remove the original producer's information and republished the video, the modified video file would have lost its integrity. In an IoT setting this could involve an unauthorized user tampering with a homeowner's smart meter to negatively impact the electricity usage readings the meter sends to the utility company. Loss of integrity does not have to be intentional, however. Visitors to the home can gain unsecured access to IoT devices and by extension impact usage data and modify settings on devices [33].

3.2.3 Challenges in Availability

As mentioned previously, IoT devices rely heavily on wireless communication. In addition to increasing a network's vulnerability to man-in-the-middle attacks, this reliance also creates a significant vulnerability to Denial of Service (DoS) attacks. An attacker could jam radio signals or time malicious message transmissions in a way that collide with legitimate transmissions. If these transmissions collide often enough, this attack could result in a negative impact to the availability of the service to the authorized user [41].

Some IoT devices are battery-powered. Through the same type of attack as above, where message transmissions collide, this can overwhelm the network and drain the battery [41].

3.3 Challenges That Are All-Encompassing in the Traditional Security Model

Other challenges do not fall neatly into the categories of the CIA Triad, but instead impact two or more of the goals of the traditional security model. The diversity of objects, for example, makes it difficult to develop adequate security measures where a one-size-fits-all approach is impossible to employ [30, 40, 43, 46, 48]. The sheer speed at which IoT devices are being adopted makes securing so many devices difficult to scale [39, 41, 42].

To match the demand for IoT devices, manufacturers are quickly generating more products, more features and more conveniences. However, manufacturers are not consistently taking the time to create standards and protocols for ensuring their devices are secure. There is no generally accepted expectation for manufacturers to create patches for their software or pre-install security measures [31, 39, 42, 46]. Plachkinova et al. discuss that manufacturers are not consistent with security-related practices such as how and where they store user information nor for how long they store it 40. Finally, regulation has also been noticeably behind the emergence of IoT devices, allowing manufacturers to get away with poor efforts to provide users with security measures [33, 45].

With the absence of manufacturer accountability, individual users are left with the responsibility of setting up their own countermeasures. However, in addition to the lack of user expertise as mentioned above, Lin et al. discuss the lack of professional services for the IoT that could fill the technical skills gap associated with providing security for these devices [46].

3.4   A Note on Privacy

It is important to note that as we began research, our assumption was that not every Security challenge is also a Privacy challenge. For example, consider a forensics investigation where a person's day-to-day activities are examined in an IoT environment. The investigator may access data, such as video footage captured as the subject of the investigation walked through her house where security cameras captured her movements. In this situation, the subject did not face a security risk. The forensics investigator was expected to review data connected to this person as part of the investigation. Privacy, on the other hand, was compromised as the investigator may have seen video footage that was not directly related to the investigation but was difficult to avoid seeing.

At a situational level, there continues to be a distinction between Security and Privacy. However, at a category level, our analysis reveals (see Fig. 1) that every Security challenge is also a Privacy challenge.

3.5   How the Challenges in Forensics, Security and Privacy Interrelate

The results of the literature review highlight 15 of the most common challenges related to forensics, security and privacy of IoT devices. These challenges are identified in Table 2 below. Furthermore, a Venn Diagram model in Fig. 1 conceptualizes how the challenges in forensics, security and privacy interrelate. This model illustrates the relationship between the challenges connected to security, forensics, and privacy with IoT devices. An in-depth analysis of our findings is further discussed in the following subsections.

To analyze the interrelation between the challenges in Forensics, Security and Privacy, each challenge category was plotted and applied to one, two or all three themes on a Venn Diagram (see Fig. 1) based on the characteristics cited in the literature review. For example, Conti et al. noted that the absence of proper authentication makes it difficult to identify responsible parties. Since this characteristic creates a challenge for both Forensics and Security, '4 – Lack of secure authentication' was plotted between Forensics and Security. Duplicates were removed after all the characteristics from the literature review were plotted. If, for example, a characteristic was plotted as causing a challenge category that applied to Security only based on a situation discussed by one author, and also plotted as causing a challenge category that applied to Forensics only based on a situation discussed by another author, Fig. 1 was updated to only include this challenge category once as overlapping between Security and Forensics to reflect the interrelation at a category level. The end result is visualized in Fig. 1.

Table 2 References categorized by forensics, security and privacy challenges

Challenge#	Challenge description	Forensics	Security	Privacy
1	Data storage not accessible	[11, 36, 56]		
2	Autonomous changes to data	[27–29]	[27, 51]	[27, 51]
3	Lack of metadata	[13, 28]		
4	Lack of secure authentication	[28, 30, 33, 36, 38, 45, 46]	[28, 30, 33, 36, 38, 40, 42, 45, 46]	[30, 33, 36, 38, 40, 42, 45, 46]
5	Unreliable big data	[1, 13, 25, 29, 37, 51]	[1]	[1, 45, 51]
6	Data in many locations	[13, 25, 29, 37, 52, 55, 56]	[13, 37]	[37, 52]
7	Difficult to acquire device	[27, 29]		
8	Diversity of objects	[25, 41–44, 46, 48, 52, 55, 56]	[41–44, 46, 48]	[41]
9	Various jurisdictions may result in legal conflict	[25, 27, 29, 45, 48, 50, 57]	[33, 48]	[25, 27, 29, 30, 51, 57]
10	No standards	[1, 23, 29, 30, 33, 36, 42, 46, 48–50, 55, 56]	[1, 29, 30, 33, 39, 42, 44, 46, 48–50]	[1, 48–50]
11	Minimal features	[1, 26, 27, 29, 30, 36, 38–40, 44, 48, 49, 55–57, 61]	[1, 30, 39, 40, 61, 48, 49, 56]	[1, 26, 30, 39, 40, 56]
12	Added security vulnerability	[1, 23, 37, 38, 41–44, 46, 48, 52, 56, 57]	[1, 23, 37, 38, 41–44, 46, 52, 56, 57]	[1, 37, 38, 56, 57]
13	Difficult to scale security measures	[1, 41]	[1, 39, 41, 42, 49, 56]	[41]
14	Lack of user expertise	[46, 49, 50]	[1, 38, 43, 44, 46, 49, 50]	[43, 44, 46, 49, 50]
15	Other	[13]	[13, 46]	[13, 46]

3.5.1 Forensics Only Challenge Categories

2 of the 15 challenge categories applied only to Forensics: 3 – Lack of metadata and 7 – Difficult to acquire device. Lack of metadata, such as time stamps for when data was last accessed, modified, or captured by IoT nodes causes a significant challenge to a forensics investigator attempting to piece together a timeline of relevant events [28]. While this characteristic creates a challenge to Forensics, it does not create a challenge to Security or Privacy because the lack of metadata functions similarly against an unauthorized user with malicious intent who is working to compromise the security or privacy of a network.

IoT nodes can be difficult to acquire as the nodes become physically smaller through manufacturing efforts to miniaturize these nodes as the potential to physically damage the nodes increases as their size decreases or because they are abnormal to a digital forensics investigation [62]. If a court refuses evidence from an alternate source, such as a home network server or cloud-based storage without having the physical device available, a forensics investigator would find it difficult to provide relevant data. Removing a physical device is more of an internal challenge that physical forensics and digital forensics departments should coordinate in the case that the device is needed by both departments for different purposes. For a criminal looking to compromise the security and/or privacy of a home network, this challenge category is likely to be irrelevant. The physical IoT node is not necessary to access the data the node can captured or that it has stored if the criminal can access the network server(s) or the cloud-based storage.

3.5.2 Security and Privacy Challenge Categories

Only 1 characteristic (within challenge category 15 – Other) applied to Security and Privacy and relates to a lack of professional services available for the IoT environment. Corporations can dedicate significant resources to staff, hardware and software that works to protect their security and privacy [46]. An IoT environment is usually built piece-by-piece without guaranteed support from each manufacturer to protect security and privacy. Anti-malware software usually places limits on the number of devices covered and, with the average number of devices per person continuously increasing, even this type of service provides limited support. This challenge understandably does not apply to Forensics. While there may be a talent gap in the number of experienced professionals in the job market, there is an infrastructure in place to have professionals available for digital forensics investigations. This infrastructure does not exist for the IoT environment.

3.5.3 Privacy and Forensics Challenge Categories

Only 1 category, 1 – Data storage not accessible, applied to Privacy and Forensics. The challenge relates specifically to 3rd party companies that control cloud-based data storage. A forensics investigation can be significantly delayed if a 3rd party company is unwilling to provide evidential data. For example, Amazon is involved in a high-profile murder case where investigators believe that an Echo device recorded evidence tied to the case [63]. The murder took place in January 2017 but Amazon has refused to release the recordings that prosecutors have requested. In November 2018 a judge ordered Amazon to release the recordings, providing relief to the forensics investigators involved in the case but not without having caused a significant delay in their ability to investigate these recordings. With a 3rd party company in control of this data, privacy is also compromised. A user does not consistently dictate what data is stored nor how long it is stored. In the case of a court order to release information, a user also loses further control of who has access to personal data.

3.5.4 Security, Privacy and Forensics Challenge Categories

The vast majority of categories (11 of 15) relate to all 3 themes: Forensics, Security and Privacy. Below are some examples of these categories and how they apply to all 3 themes.

Category 2 – Autonomous changes to data, means that IoT nodes capturing data real-time make it difficult for a forensics investigator to preserve information that shows a clear timeline of events [58]. Instead this autonomously changing node captures a snapshot that may not include all relevant details. Also, because data is changed without human input, it's difficult to identify who is responsible for a security or privacy breach [62].

Category 4 – Lack of secure authentication makes it difficult for a forensics investigator to identify responsible parties [58]. This also leaves a user vulnerable to unauthorized access and use of her IoT network.

Category 5 – Unreliable big data causes a challenge to Forensics due to the large amount of data to analyze [25]. An individual user would likewise find it challenging to analyze the large amount of data available to determine the details of a security or privacy breach of his IoT network.

Category 9 – Various jurisdictions may result in legal conflict creates a number of challenges to Forensics. For example, if a crime takes place in one country but evidential data is stored in cloud-based storage owned by a company in another country, the two jurisdictions will need time to decide how to handle the needs of the investigation, potentially causing significant delays to the investigation [29]. Depending on the jurisdiction, the privacy of the user may or may not be taken into consideration as part of the decision-making process. Seen from a different angle, this category also creates challenges based on the lack of legal authorities keeping manufacturers accountable to taking measures to protect user security and privacy. Not only are there insufficient legal authorities, the various jurisdictions make what legal efforts do exist lack unified policies [33].

4 Case Study

4.1 Drone Ubiquity

Drones have taken much of the world by surprise and by storm; they have even penetrated and rejuvenated stagnant industries that are not well known for advancements in innovation. There is a consensus from the articles reviewed for this chapter that drones are rapidly becoming a part of everyday life and will continue trending towards ubiquitous usage as their level of technology, capabilities, and ease of use improves.

Drones have already gone through many iterations of technology generational capabilities; below is a list of the technology generations [34, 64]:

- Generation 1: Basic remote-control aircraft of all forms
- Generation 2: Static design, fixed camera mount, video recording and still photos, manual piloting control
- Generation 3: Static design, two-axis gimbals, HD video, basic safety models, assisted piloting
- Generation 4: Transformative designs, Three-axis gimbals, 1080P HD video or higher-value instrumentation, improved safety modes, autopilot modes.
- Generation 5: Transformative designs, 360° gimbals, 4 K video or higher-value instrumentation, intelligent piloting modes.
- Generation 6: Commercial suitability, safety and regulatory standards-based design, platform and payload adaptability, automated safety modes, intelligent piloting models and full autonomy, airspace awareness
- Generation 7: Complete commercial suitability, fully compliant safety and regulatory standards-based design, platform and payload interchangeability, automated safety modes, enhanced intelligent piloting models and full autonomy, full airspace awareness, auto action (takeoff, land, and mission execution)

Taking a note from the innovation of companies such as Amazon and their plan to utilize drones as expedient delivery tool, criminals have started using drones for smuggling contraband into prisons along with transporting drug packages across international borders. This can create a major problem for authorities when trying to investigate and interdict such activities as the pilots of the drones can easily get away with little fear of being captured. Many modern drones have the ability to encrypt the data that investigators would need in order to track them down. Forensics provide essential assistance in retrieving and deciphering this data to assist in catching these criminals. Even when deciphered by forensics, it can still be very difficult to catch the criminals. Another utilization for drones, on the more extreme end, is using them as weapons for warfighting and assassinations. Drones have been heavily utilized by militaries of the world, from superpowers to third-world country freedom fighters. They have even recently been used in an attempted assassination in Venezuela. However, military usage is still by far the most common in the weapons category.

A very interesting future-concept of drone usage as weapons has been presented by the Convention of Conventional Weapons; it shows the use of autonomous drones, via advanced AI, that can be programmed to attack certain types of targets automatically, such as specific categories of people for instance [65, 66]. They can be dropped from military aircraft by the hundreds to thousands via the use of "mothership" drones that carry four to eight miniature attack drones. The mothership drone has a greater distance capability and deploys the short-range attack units when it detects the enemy is within range. The concept design demonstrated attack drones with small shaped-charge explosives that seek out the predefined target criteria then hone in to the head of the target and detonate, killing them. This demonstration also pointed out what could happen when this technology falls into the wrong hands and is hacked and reprogrammed to do their bidding. The "bad guys" of the video steal the military drones and reprogram them to attack students at a university who belong to an opposing political party. It is not clearly stated how the drone knew the political agenda of the target; assumable that it could use some form of face-recognition and seek out pre-assigned targets based upon positive facial matches. The point is very poignant however, you could simply program the drone to attack people of a certain skin color for instance. If such an occurrence were to ever happen, forensics would be crucial in finding who was responsible.

With so many potential uses for drones to do horrible and unspeakable acts, the ability to conduct forensic analysis on them is crucial.

4.2 Drone Forensics

Drones are still relatively new to the scene overall. However, drones are not new or an unknown entity to forensics as they are an amalgam of existing technologies that already have well-established forensic tools and can be broken down into component parts such as Internet of Things (IoT), Cyber Physical Engineering System (CPES), a network of sensors, storage, CPUs and actuators with network connections that enable them to share data and control information. All of these components are involved in a complex, often real-time, flow of telemetry, sensor, and environmental data in clear text and encrypted formats [55].

The main difference drone forensics have from other forms of digital forensics is in the type of data that is collected and what portion of that data is of importance/interest. It is interesting to note that a drone itself is an instance of IoT/CPES along with one to many CPUs, networked sensors that communicate with various external systems. Some external systems that the drone may be communicating with are part of a larger Unmanned Aerial System or UAS. The UAS will usually consist of components such as a remote controller, a Ground Control Station (GCS), and a mobile device such as a cell phone or tablet. To properly conduct forensics that produce meaningful and useful data, drones must be broken down into their component parts. Much of the recoverable data and metadata will be more useful when retrieved from the cooperating component parts. Often this data will prove to be useless to forensic efforts without the data from the other component parts.

The following data types are potentially recoverable from drones and their control devices/systems:

Owner Indemnifying Information
Drone Name
Drone Serial Number
Launch Point Location
WIFI and/or Bluetooth data
Flight Data (GPS coordinates, Waypoints, etc.)
Sensor Logs
Recorded Media (Videos, Images, EXIF data, etc.)
Mobile Phone Data (if controlled by one)
Uploaded Data
Cloud Data
Social Media Data
Operation System Data
Ground Controls System Data

Some of the most popular drones today are manufactured by DJI, one of the largest consumer drone manufacturers from China. DJI drones have even been used by some militaries around the world. There is a forensic tool designed specifically for DJI Phantom Drones that is available called the Drone Open source Parser (DROP). DROP has the ability to parse proprietary DAT files extracted from DJI's phantom drones. These DAT files are encrypted and encoded to prevent unauthorized access. DROP can also extract TXT files containing highly useful information such as GPS location(s), battery level(s), flight time(s), and other metadata that can be correlated to control device data and identify a match between drone and controller based on this recovered data [63]. This works well for mainstream manufactured drones. However, there are many technologies and methods to make drone forensics much more difficult such as the ability to 3D print and/or build a custom-made drone with widely available component parts that are device-agnostic that can be programmed from freely available software repositories, potentially making the forensics process much more difficult.

Below are potential innovations and/or existing technical difficulties that can confound the forensic process on drones:

Many drone manufacturers beyond the mainstream companies such as DJI, not all conform to a standard.
Drones are part of a UAS, all components of a UAS, such as the drone itself cell phone or tablet and remote controller, will need to be analyzed to obtain meaningful and useful data.
Artificial Intelligence (AI) and machine learning to control drone swarms autonomously

3D printing technology will enable custom, one-off/one-time-use, significantly modified drones that do not conform to standards confounding forensic techniques.

The technology that powers drones is constantly evolving giving drones more complex capabilities, thus making forensic analysis more involved and/or difficult.

4.3 Drones as a Forensic Tool

There is a growing trend in utilizing drones as forensic tools themselves, taking advantage of the unique capabilities that drones bring to the table. One of the most interesting cases reviewed for this type of use is by the company **Persistent Surveillance Systems**. A podcast put on by RadioLab, in which they interviewed the founder and CEO of Persistent Surveillance Systems Ross McNutt, went into great detail on what their business model provides via the use of drones. A drone is put into the sky at approximately 18,000 ft. and flies over a target area, a city for example, continuously for days as it snaps a 44 megapixel image of the target area every second and transmits it back to their headquarters for processing [66, 67]. When an action of interest occurs, a crime for example, the footage is reviewed at the scene and time of the crime and then further reviewed backwards as far as they have record to witness what happens in that area until they can determine what events led to the scene of the crime. For example, a police officer was assassinated in Juarez Mexico while she stopped at a traffic light; this system was used to identify the vehicle that the assassins used, and its path was traced back until accomplices were identified, all the way back to the location of the crime lord's base of operations. This crime lord was said to be responsible for thousands of similar assassinations. All of this was made possible by simply reviewing the history of video events 1 s snapshot at a time [68]. The cameras at the time of this incident were 44 megapixel cameras, they now employ drones that have 192 megapixel, full color, camera systems found in their HawkEye II [69] (see Fig. 2).

This technology was originally developed by Ross McNutt and a team while still in the US Military. It was developed out of a necessity to see when, where and how Improved Explosive Devices (IEDs) were placed as traps. The system proved very effective at identifying, locating and eventually leading to the apprehension of the combatants who placed the IEDs.

5 Conclusion

The proliferation of the Internet of Things (IoT) has added many benefits to our lives. However, it also brings with it many vulnerabilities, as described in the Drone case study. The growing number of security breaches far outweigh the slow development of solutions for IoT devices. It is paramount to public safety and

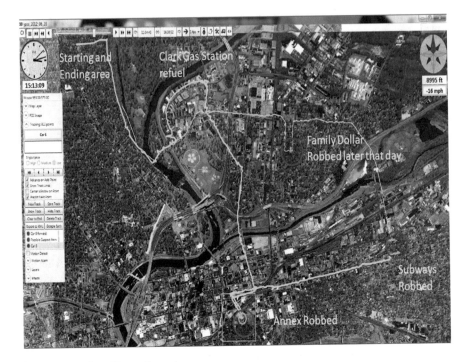

Fig. 2 Drone Surveillance Forensics

crime prevention/solving that law enforcement, corrections, security and military professionals have the knowledge, tools and level of preparedness that allows them to conduct meaningful and successful forensic analysis on IoT devices such as drones and their component systems.

By identifying the interrelatedness of the challenges in forensics, security and privacy, this chapter encourages the design of safer and more secure IoT technology to prevent misuse. It is important that policy makers, manufacturers and other stakeholders involved direct their efforts to develop effective solutions to the current challenges of IoT devices.

5.1 Future Research

While the number of articles reviewed show that a significant effort is being placed in research on Forensics, Security and Privacy in an IoT environment and, on a much broader level, for the Internet of Things, many authors echo that the research is still in a stage of infancy. The opportunities for future research have the underlying theme of prioritizing and improving strategies for Forensics, Security and Privacy.

5.1.1 Investigation of the Most Frequently Mentioned Challenge Categories

Table 1 illustrates the challenge categories in the context of one or more of the 3 themes: Forensics, Security and/or Privacy. There is a clear and visible concentration of other authors discussing topics such as category 10 – No standards, with 15 articles exploring this category. Category 3 – Lack of metadata, on the other hand, was only discussed in 2 of the articles reviewed. Prioritizing efforts on the effects that no standards have on the IoT environment could yield much more powerful results than prioritizing efforts on the lack of metadata. Another area of research would be to analyze how the top 3 challenges mentioned most frequently by the authors in the literature review: 'No standards', 'Minimal features' and 'Added security vulnerability', contribute to IoT being possible attack vectors.

5.1.2 Investigation of the Effect of One Challenge Category on Another Category

Research efforts could benefit from a domino effect if these efforts are focused on challenge categories that affect one or more other challenge categories. For example, work directed to address challenge category 10 – No standards, would likely also mitigate challenge categories such as 4 – Lack of secure authentication, 6 – Data in many locations, 8 – Diversity of objects and 13 – Difficult to scale security measures. For example, if manufacturers took accountability to setting standards in pre-installing intrusion detection software, this standard would reduce the impact on the diverse types of IoT devices in a IoT environment and the work done by the manufacturers would help scale the security measures rather than individual users or organizations addressing the implementation of a scalable solution.

5.1.3 Investigation of the Proposed Solutions to the Challenges of IoT Devices

This work can further be expanded by examining current trends that seek to resolve the vulnerabilities of IoT devices and perform a comparative analysis of the relationship between the solutions and the challenges. Our preliminary literature review supports this perspective. According to Khan, standards, protocols, trust architectures and a global body for regulating IoT should be established to mitigate the current challenges highlighted in this chapter. Gubbi proposes that the future of IoT involve a scalable cloud framework that includes network, computation, storage and data visualization [47]. Including both challenges and proposed solutions could provide more insight and knowledge on how to better protect IoT users.

References

1. B.B. Zarpelão, R.S. Miani, C.T. Kawakani, S.C. de Alvarenga, A survey of intrusion detection in Internet of Things. J. Netw. Comput. Appl. **84**, 25–37 (2017). https://doi.org/10.1016/j.jnca.2017.02.009
2. A. Yazdinejad, R.M. Parizi, A. Dehghantanha, K.-K.R. Choo, Blockchain-enabled authentication handover with efficient privacy protection in SDN-based 5G networks. IEEE Trans. Netw. Sci. Eng. (2019). https://doi.org/10.1109/TNSE.2019.2937481
3. A. Azmoodeh, A. Dehghantanha, K.-K.R. Choo, Robust malware detection for internet of (battlefield) things devices using deep eigenspace learning. IEEE Trans. Sustain. Comput. **4**(1), 88–95 (2018)
4. M. Saharkhizan, A. Azmoodeh, H. HaddadPajouh, A. Dehghantanha, R.M. Parizi, G. Srivastava, A hybrid deep generative local metric learning method for intrusion detection, in *Handbook of Big Data Privacy*, (Springer, 2020), pp. 343–357. https://doi.org/10.1007/978-3-030-38557-6_16
5. M.M. BehradFar et al., RAT hunter: Building robust models for detecting remote access trojans based on optimum hybrid features, in *Handbook of Big Data Privacy*, (Springer, 2020), pp. 371–383. https://doi.org/10.1007/978-3-030-38557-6_18
6. M. Riggins, Ubiquitous: At what costs? (Inspired eLearning, 2017), https://inspiredelearning.com/wpcontent/uploads/2017/05/Ubiquitous_At-What-Cost.pdf
7. A. Yazdinejad, R.M. Parizi, A. Dehghantanha, K.-K.R. Choo, P4-to-blockchain: A secure blockchain-enabled packet parser for software defined networking. Comput. Secur. **88** (2020). https://doi.org/10.1016/j.cose.2019.101629
8. I. Hamilton, A judge has ordered Amazon to hand over recordings from an Echo to help solve a double murder case, Business Insider, 2018, https://www.businessinsider.in/a-judge-has-ordered-amazon-tohand-over-recordings-from-an-echo-to-help-solve-a-doub, p. 66605145 (2020)
9. A. Azmoodeh, A. Dehghantanha, K.-K.R. Choo, Big data and internet of things security and forensics: Challenges and opportunities, in *Handbook of Big Data and IoT Security*, (Springer, 2019), pp. 1–4. https://doi.org/10.1007/978-3-030-10543-3_1
10. H. HaddadPajouh, R. Khayami, A. Dehghantanha, K.-K.R. Choo, R.M. Parizi, AI4SAFE-IoT: An AI-powered secure architecture for edge layer of Internet of things. Neural Comput. Applic. **32**(20), 16119–16133 (2020). https://doi.org/10.1007/s00521-020-04772-3
11. S. Watson, A. Dehghantanha, Digital forensics: The missing piece of the Internet of Things promise. Comput. Fraud Secur. **2016**(6), 5–8 (2016). https://doi.org/10.1016/s1361-3723(15)30045-2
12. M. E. and R. Enright, D. Palmer, N. Dawson, Samsung open economy report
13. E. Casey, The value of forensic preparedness and digital-identification expertise in smart society. Digit. Investig. **22**, 1–2 (2017). https://doi.org/10.1016/j.diin.2017.09.001
14. A. Aminnezhad, A. Dehghantanha, M.T. Abdullah, A survey on privacy issues in digital forensics. Int. J. Cyber-Security Digit. Forensics **1**(4), 311–324 (2012)
15. K.-K.R.C.A. Dehghantanha, Eda, *Handbook of Big Data Privacy* (Springer, Cham, 2020)
16. A. Yazdinejad, G. Srivastava, R.M. Parizi, A. Dehghantanha, H. Karimipour, S.R. Karizno, SLPoW: Secure and low latency proof of work protocol for blockchain in green IoT networks, in *2020 IEEE 91st Vehicular Technology Conference (VTC2020-Spring)*, (2020), pp. 1–5
17. A. Singh, K. Click, R.M. Parizi, Q. Zhang, A. Dehghantanha, K.-K.R. Choo, Sidechain technologies in blockchain networks: An examination and state-of-the-art review. J. Netw. Comput. Appl. **149**, 102471 (2020). https://doi.org/10.1016/j.jnca.2019.102471
18. A. Yazdinejad, R.M. Parizi, A. Dehghantanha, Q. Zhang, K.-K.R. Choo, An energy-efficient SDN controller architecture for IoT networks with blockchain-based security. IEEE Trans. Serv. Comput. (2020). https://doi.org/10.1109/TSC.2020.2966970
19. D. Połap, G. Srivastava, A. Jolfaei, R.M. Parizi, Blockchain technology and neural networks for the internet of medical things, in *IEEE INFOCOM 2020 – IEEE Conference on Computer*

Communications Workshops (INFOCOM WKSHPS), (2020), pp. 508–513. https://doi.org/10.1109/INFOCOMWKSHPS50562.2020.9162735

20. A. Yazdinejad, G. Srivastava, R.M. Parizi, A. Dehghantanha, K.-K.R. Choo, M. Aledhari, Decentralized authentication of distributed patients in hospital networks using blockchain. IEEE J. Biomed. Health Inform. **24**(8), 2146–2156 (2020)

21. Q. Chen, G. Srivastava, R.M. Parizi, M. Aloqaily, I. Al Ridhawi, An incentive-aware blockchain-based solution for internet of fake media things. Inf. Process. Manag., 102370 (2020). https://doi.org/10.1016/j.ipm.2020.102370

22. A. Yazdinejad, R.M. Parizi, A. Bohlooli, A. Dehghantanha, K.-K.R. Choo, A high-performance framework for a network programmable packet processor using P4 and FPGA. J. Netw. Comput. Appl. **156**, 102564 (2020)

23. A. Bujari, M. Furini, F. Mandreoli, R. Martoglia, M. Montangero, D. Ronzani, Standards, security and business models: Key challenges for the IoT scenario. Mob. Netw. Appl. **23**(1), 147–154 (2018). https://doi.org/10.1007/s11036-017-0835-8

24. S. Khan, M. Shiraz, A.W.A. Wahab, A. Gani, Q. Han, Z.B.A. Rahman, A comprehensive review on adaptability of network forensics frameworks for mobile cloud computing. ScientificWorldJournal **2014**, 547062 (2014). https://doi.org/10.1155/2014/547062

25. Q. Do, B. Martini, K.-K.R. Choo, Cyber-physical systems information gathering: A smart home case study. Comput. Netw. **138**, 1–12 (2018). https://doi.org/10.1016/j.comnet.2018.03.024

26. S. Perumal, N.M. Norwawi, V. Raman, Internet of Things(IoT) digital forensic investigation model: Top-down forensic approach methodology, in *2015 Fifth International Conference on Digital Information Processing and Communications (ICDIPC)*, (IEEE, 2015). https://doi.org/10.1109/icdipc.2015.7323000

27. E. Oriwoh, G. Williams, *Internet of Things: The Argument for Smart Forensics. Handbook of Research on Digital Crime, Cyberspace Security, and Information Assurance* (IGI-Global Publishing, 2014)

28. A. Awasthi, H.O.L. Read, K. Xynos, I. Sutherland, Welcome PWN: Almond smart home hub forensics. Digit. Investig. **26**, S38–S46 (2018). https://doi.org/10.1016/j.diin.2018.04.014

29. E. Kalaimannan, Smart device forensics – Acquisition, analysis and interpretation of digital evidences, in *2015 International Conference on Computational Science and Computational Intelligence (CSCI)*, (IEEE, 2015). https://doi.org/10.1109/csci.2015.58

30. L. Babun, A.K. Sikder, A. Acar, A.S. Uluagac, Iotdots: A digital forensics framework for smart environments. arXiv Prepr. arXiv1809.00745 (2018)

31. B.L.R. Stojkoska, K.V. Trivodaliev, A review of Internet of Things for smart home: Challenges and solutions. J. Clean. Prod. **140**, 1454–1464 (2017)

32. T. Zia, P. Liu, W. Han, Application-specific digital forensics investigative model in Internet of Things (IoT), in *Proceedings of the 12th International Conference on Availability, Reliability and Security*, (ACM, 2017). https://doi.org/10.1145/3098954.3104052

33. M. Plachkinova, A. Vo, A. Alluhaidan, Emerging trends in smart home security, privacy, and digital forensics, in *Proceedings of the 22nd Americas Conference on Information Systems*, (2016)

34. A. Yazdinejad, R.M. Parizi, A. Dehghantanha, H. Karimipour, G. Srivastava, M. Aledhari, Enabling drones in the internet of things with decentralized blockchain-based security. IEEE Internet Things J., 1 (2020). https://doi.org/10.1109/jiot.2020.3015382

35. Z.A. Baig et al., Future challenges for smart cities: Cyber-security and digital forensics. Digit. Investig. **22**, 3–13 (2017). https://doi.org/10.1016/j.diin.2017.06.015

36. Q. Do, B. Martini, K.-K.R. Choo, *FAU Open Research Challenge: Digital Forensics – Forensic Report* (2015), p. 17

37. E. Oriwoh et al., A comprehensive review on adaptability of network forensics frameworks for mobile cloud computing. J. Clean. Prod. **140**(1), 1–12 (2018). https://doi.org/10.1145/3098954.3104052

38. M. Schiefer, Smart home definition and security threats, in *2015 Ninth International Conference on IT Security Incident Management & IT Forensics*, (IEEE, 2015). https://doi.org/10.1109/imf.2015.17

39. S. Sicari, A. Rizzardi, L.A. Grieco, A. Coen-Porisini, Security, privacy and trust in Internet of Things: The road ahead. Comput. Netw. **76**, 146–164 (2015). https://doi.org/10.1016/j.comnet.2014.11.008
40. M.O.F.A. Alaba, Internet of things security: A survey (Scribd, 2017). Available [Online]: https://www.scribd.com/document/360075916/Alaba-Othman-et-al-2017-Internet-of-things-Survey-pdf. Accessed 16 Sep 2020
41. Q.M. Ashraf, M.H. Habaebi, Autonomic schemes for threat mitigation in Internet of Things. J. Netw. Comput. Appl. **49**, 112–127 (2015). https://doi.org/10.1016/j.jnca.2014.11.011
42. Z.-K. Zhang, M.C.Y. Cho, C.-W. Wang, C.-W. Hsu, C.-K. Chen, S. Shieh, IoT security: Ongoing challenges and research opportunities, in *2014 IEEE 7th International Conference on Service-Oriented Computing and Applications*, (IEEE, 2014). https://doi.org/10.1109/soca.2014.58
43. K. Sha, W. Wei, T. Andrew Yang, Z. Wang, W. Shi, On security challenges and open issues in Internet of Things. Futur. Gener. Comput. Syst. **83**, 326–337 (2018). https://doi.org/10.1016/j.future.2018.01.059
44. R. Heartfield et al., A taxonomy of cyber-physical threats and impact in the smart home. Comput. Secur. **78**, 398–428 (2018). https://doi.org/10.1016/j.cose.2018.07.011
45. N. Komninos, E. Philippou, A. Pitsillides, Survey in smart grid and smart home security: Issues, challenges and countermeasures. IEEE Commun. Surv. Tutorials **16**(4), 1933–1954 (2014). https://doi.org/10.1109/comst.2014.2320093
46. H. Lin, N. Bergmann, IoT privacy and security challenges for smart home environments. Information **7**(3), 44 (2016). https://doi.org/10.3390/info7030044
47. D. Geneiatakis, I. Kounelis, R. Neisse, I. Nai-Fovino, G. Steri, G. Baldini, Security and privacy issues for an IoT based smart home, in *2017 40th International Convention on Information and Communication Technology, Electronics and Microelectronics (MIPRO)*, (IEEE, 2017). https://doi.org/10.23919/mipro.2017.7973622
48. S. Notra, M. Siddiqi, H. Habibi Gharakheili, V. Sivaraman, R. Boreli, An experimental study of security and privacy risks with emerging household appliances, in *2014 IEEE Conference on Communications and Network Security*, (IEEE, 2014). https://doi.org/10.1109/cns.2014.6997469
49. Z. Whittaker, After massive cyberattack, shoddy smart device security comes back to haunt | ZDNet (2016), https://www.zdnet.com/article/blame-the-internet-of-things-for-causing-massive-web-outage/. Accessed 16 Sept 2020
50. E. Ronen, A. Shamir, Extended functionality attacks on IoT devices: The case of smart lights, in *2016 IEEE European Symposium on Security and Privacy (EuroS&P)*, (IEEE, 2016). https://doi.org/10.1109/eurosp.2016.13
51. J. Lopez, R. Rios, F. Bao, G. Wang, Evolving privacy: From sensors to the Internet of Things. Futur. Gener. Comput. Syst. **75**, 46–57 (2017). https://doi.org/10.1016/j.future.2017.04.045
52. R. McKemmish, *What Is Forensic Computing?* (Australian Institute of Criminology, Canberra, 1999)
53. A. Yazdinejad, A. Bohlooli, K. Jamshidi, Efficient design and hardware implementation of the OpenFlow v1.3 Switch on the Virtex-6 FPGA ML605. J. Supercomput. **74**(3) (2018). https://doi.org/10.1007/s11227-017-2175-7
54. A. Yazdinejad, R.M. Parizi, G. Srivastava, A. Dehghantanha, K.-K.R. Choo, Energy efficient decentralized authentication in internet of underwater things using blockchain, in *2019 IEEE Globecom Workshops (GC Wkshps)*, (2019), pp. 1–6
55. M. Banerjee, J. Lee, K.K.R. Choo, A blockchain future to Internet of Things security: A position paper (Digital Communications and Networks, 2017), http://www.Sci.com/science/article/piiS
56. J. Gubbi, R. Buyya, S. Marusic, M. Palaniswami, Internet of Things (IoT): A vision, architectural elements, and future directions. Futur. Gener. Comput. Syst. **29**(7), 1645–1660 (2013). https://doi.org/10.1016/j.future.2013.01.010
57. V.S. Harichandran, F. Breitinger, I. Baggili, A. Marrington, A cyber forensics needs analysis survey: Revisiting the domain's needs a decade later. Comput. Secur. **57**, 1–13 (2016). https://doi.org/10.1016/j.cose.2015.10.007

58. M. Conti, A. Dehghantanha, K. Franke, S. Watson, Internet of Things security and forensics: Challenges and opportunities. Futur. Gener. Comput. Syst. **78**, 544–546 (2018). https://doi.org/10.1016/j.future.2017.07.060

59. K. Bolouri, A. Azmoodeh, A. Dehghantanha, M. Firouzmand, Internet of Things camera identification algorithm based on sensor pattern noise using color filter array and wavelet transform, in *Handbook of Big Data and IoT Security*, (Springer, 2019), pp. 211–223. https://doi.org/10.1007/978-3-030-10543-3_9

60. H.M. Rouzbahani, Z. Faraji, M. Amiri-Zarandi, H. Karimipour, AI-enabled security monitoring in smart cyber physical grids, in *Security of Cyber-Physical Systems*, (Springer, Cham, 2020), pp. 145–167. https://doi.org/10.1007/978-3-030-45541-5_8

61. F. Murtagh, Big data scaling through metric mapping, in *Data Science Foundations*, (Chapman and Hall/CRC, 2017), pp. 103–129. https://doi.org/10.1201/9781315367491-6

62. A. Azmoodeh, A. Dehghantanha, M. Conti, K.-K.R. Choo, Detecting crypto-ransomware in IoT networks based on energy consumption footprint. J. Ambient. Intell. Humaniz. Comput. **9**(4), 1141–1152 (2018)

63. D.R. Clark, C. Meffert, I. Baggili, F. Breitinger, DROP (DRone Open source Parser) your drone: Forensic analysis of the DJI Phantom III. Digit. Investig. **22**, S3–S14 (2017). https://doi.org/10.1016/j.diin.2017.06.013

64. D. Joshi, Drone technology uses and applications for commercial, industrial and military drones in 2020 and the future, Bus. Insid., 2019

65. StratoEnergetics, Slaughterbots, 2017

66. C. Timberg, New surveillance technology can track everyone in an area for several hours at a time, Washington Post, 2014

67. D. Kovar, J. Bollo, Drone forensics. Digital Forensics Magazine **34**, 7–2018

68. WNYC Studios, Radiolab: Eye in the sky (WNYC Studios, 2015), https://www.wnycstudios.org/podcasts/radiolab/articles/eye-sky. Accessed 16 Sept 2020

69. Persistent Surveillance Systems, Persistent surveillance systems (2016), https://www.pss-1.com/. Accessed 16 Sept 2020

Detection of Enumeration Attacks in Cloud Environments Using Infrastructure Log Data

Samira Eisaloo Gharghasheh and Tim Steinbach

1 Introduction

During the past decade utilizing cloud-based environments has experienced a significant increase in terms of diversity and scalability and this increase has encouraged cybercriminals to target cloud-based environments [1–5]. Enumeration refers to the process of extracting resources, machine names, and services with setting an active connection to the target [6–9]. Considering their popularity and potential to save organizational resources, reduce their cost and make them more flexible, cloud environments have become an attractive target for cyber-criminals, especially in enumeration activities [10–14]. Presently, we need more advanced and algorithmic techniques such as machine learning and deep learning to detect cyber threats. One of the most appropriate attack detection and identification techniques in cloud infrastructure is based on log analysis [15, 16]. Because cloud log files are huge in size, using threat hunting techniques without getting help from machine learning and deep learning algorithms could be time-consuming. Deep Neural Networks (DNNs) [17] are capable of mapping nonlinear functions [18]. Recurrent Neural Networks (RNNs) as a type of DNNs are used to add context into word vectors. By far the most popular RNNs are LSTMs [19–22]. They are beneficial in Natural Language Processing (NLP) and can help in log analysis considering log entries as text. LSTMs owing to the fact that they have memory, they can remember the calculated value so far [21].

S. E. Gharghasheh (✉)
School of Computer Science, University of Guelph, Guelph, ON, Canada
e-mail: samira@cybersciencelab.org

T. Steinbach
eSentire Inc., Waterloo, ON, Canada
e-mail: tim.steinbach@esentire.com

© Springer Nature Switzerland AG 2021
K.-K. R. Choo, A. Dehghantanha (eds.), *Handbook of Big Data Analytics and Forensics*, https://doi.org/10.1007/978-3-030-74753-4_3

In this paper, Long Short-Term Memory (LSTM) and Convolutional Neural Network (CNN) was used to detect enumeration attacks on Active Amazon Web Services (AWS) log data. We had used many to one relationship to classify log entries. The inputs were sequences of features of each log entries and the output were a dummy variable of 0 indicating non-malicious or 1 indicating malicious. To access attack logs in AWS, a number of scripts had been run against it using Application Programming Interface (API) key. A script was implemented toward parsing the JSON format CloudTrail log data to CSV. Concerning the huge log data size, labeling the data manually was challenging. Hence, to overcome this issue, a labeling script was implemented with the help of some sort of time windowing the data and deciding whether it is a malicious activity or not by the number of activities which had been done by a single user or an IP address in that time window and corresponding error messages occurred by that time. For evaluation measures we used accuracy, loss, validation accuracy and validation loss.

A basic dashboard was developed to report found malicious logs and categorize them by their running hours per day, the most called event names, top suspicious roles, and top suspicious IP addresses. These features assist users to better monitor, measure and be aware of threats [23].

Section 2 of this paper presents a literature review of the related works from recent years. Section 3 contains the methodology and Sect. 4 illustrates the visualization dashboard. Section 5 provides the results and in Sect. 6, a conclusion is drawn.

2   Related Works

The authors in [24] used Support Vector Machines (SVM), Decision Tree, Neural Network and Random Forest for detecting network threats on cloud computing and achieved an accuracy of 95%, 96%, 98% and 99% respectively. Researchers in [25] benefited from the CNN models to detect cloud-based attacks and achieved an accuracy of 79.93% in attack detection. In [26], they used Linear Regression (LR) and Random Forest (RF) for the purpose of anomaly detection in multi-cloud environments. They achieved 99% detection accuracy and 93.6% categorization accuracy. To classify cache-based applications in a cloud environment, [27] built a deep learning model that achieved a classification rate of 98%. Ten machine learning algorithms were used in [28] to classify web pages into Cross-Site Scripting (XSS) and non-XSS categories. Their evaluation showed better performance in the Social Networking Services (SNS) with the accuracy of 97.2% and False Positive Rate (FPR) of 87%. A deep learning framework was proposed in [29] to detect cyber-attacks in mobile cloud computing environments. They attained the highest accuracy of 97.11% in attack detection. Machine learning algorithms were used in [30] for anomaly detection and to defend against Advanced Persistent Threat (APT) actors' activities. Among the supervised machine learning algorithms used in this paper, the SVM achieved the True Positive Rate (TPR) of 95.33% and one-class SVMs achieved the TPR of 98.67%. The SVM algorithms were trained based on the normal behaviour of the users. Moreover, in [31] they used SVM algorithms and achieved 87.8% accuracy of detecting intrusions at the network layer. In the aforesaid paper, the detection rate accuracy reached almost 100% with a false alarm rate of 2.8%. An APT Unsupervised Learning Detection (AULD) system was proposed in [32] to detect suspicious domains using unsupervised machine learning algorithms. They used Domain Name Server (DNS) log data, to extract important features. Then they listed 1,584,225,274 DNS records among their most suspicious data.

Their proposed AULD successfully detected all the domain names used in APT attacks. In [33], the researchers proposed a real-time intrusion detection system using system logs. Azmoodeh et al. [34] presented a ransomware detection method that analyzed log information of internet of thing devices and detects attacks based on process energy consumptions within logfiles. The mentioned system was able to categorize the log based on their severity to three levels of high, medium, and low. Furthermore, their system raised an alert when it detected any abnormal behaviour. In order to detect attacks based on DNS requests, [35] proposed a new deep learning-based method. They archived an accuracy of 97.6% with a false positive rate of 2.3%. It is noteworthy that [36] put forward a deep learning method to discover network scanning activities from Apache HTTP server access logs. They obtained a 99.38% accuracy and 100% precision rate. Saharkhizan et al. [37] presented a deep generative model that learns complex structure of network attack traffics for detecting anomalies and obtained high detection rate and low false alarm. A Distributed Denial of Service (DDoS) attack detection system using the C.4.5 algorithm was proposed in [38]. They used signature detection-based techniques to enhance attack detection. Researchers in [39] used deep learning to detect anomalies from system logs and they called it DeepLog. They achieved an accuracy of almost 100% in anomaly detection. An anomaly detection model based on deep neural network structures proposed in [40]. They evaluated their models on a test dataset and the highest area under Receiver Operating Characteristic (ROC) curve among the tested algorithms belongs to Deep Convolutional Neural Network (DCNN) and was 0.955. For the purpose of anomaly detection in system logs, [41] presented the RNN model and their approach on the receiver operator characteristic curve was 0.99 [42]. performed a deep learning anomaly detection based on the image completion model. Their proposed algorithm attained the area under the ROC curve of 0.95. There is limited research in detecting attacker's enumeration and reconnaissance activities from cloud infrastructure logs, a challenge that will be addressed in this research.

3   Methodology

The proposed approach consists of five steps which is demonstrated in Fig. 1. First section describes the process of running an enumeration attack on AWS to have malicious log entries for training models. Section two describes the CloudTrail log

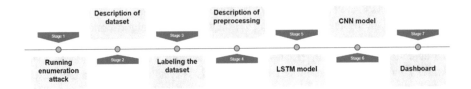

Fig. 1 An overview of the research

and parsing script. Then, the preprocessing of the dataset is explained which is subsequently followed by a presentation of the LSTM model as the solution. Finally, CNN is introduced as another deep neural network solution.

3.1 Running Enumeration Attack

In the interest of having malicious log entries in the type of enumeration attack logs, different attack scripts that adopted from [43, 44] were run against AWS. For having various IP addresses, multiple LightSail machines had been used to run attacks with different roles and credentials. Instead of just calling the top 10 AWS services, we were able to enumerate through each feature for all services with the help of these tools.

3.2 Description of Dataset

The dataset which was downloaded from CloudTrail, used to train, and test the created model that includes 89,000 malicious and 361,000 non-malicious log entries. The script that we used for downloading log data was also used for parsing the JSON format log file to the CSV format. Features of the dataset are described in Table 1.

3.3 Labeling the Dataset

To label the huge dataset, we implemented a script which made use of various rules that were adopted from eSentire's [45] specialists. For the first step, logs were divided into time windows. Then, the log entries in each time window were grouped by their IP addresses. The decision for the log entry is malicious or non-malicious made by the following rules. If one source IP is making 5 different get, list, and describe calls and there are errors on more than 60% of the calls, that might be malicious. A similar rule was added to the script for roles instead of IP addresses.

Table 1 Feature description

No. feature	Feature name	Description
1	userIdentity_type	Type of user who made the call
2	userIdentity_arn	Detail information about that user
3	userIdentity_sessionContext_sessionIssuer_type	Session issuer type
4	userIdentity_sessionContext_attributes_creationDate	Session creation date
5	userIdentity_userName	Username
6	eventTime	The time of the event
7	eventSource	The source for the event
8	eventName	Call types on different resources
9	awsRegion	AWS region
10	sourceIPAddress	IP address which made the call
11	userAgent	The user agent which made the call
12	errorCode	Error type which occurred during the call
13	eventType	AWS event type

3.4 Description of Preprocessing

Data preprocessing is a critical operation due to the impacts it has on the victory of every data mining model. For the purposes of preprocessing in this project, a list of actions was done which are listed as follows:

- Instead of a creation date for each log entry, we used 1 or 0 whether it had a session date or not respectively.
- Features with unique values such as event time and source IP addresses were dropped.
- Fill missing values with NaN.
- Shuffle the dataset to have a representative of the overall distribution of the data in each training, test, and validation portions.
- Tokenizing the dataset and taking 2000 common words.
- Using fit_on_text from Keras to go through all the data and create a dictionary and index words by the most used through all the data.
- With the help of texts_to_sequences from Keras, all tokens from the previous step turned to sequences.
- To have the same size of sequences for better training the data, padding was used.

3.5   LSTM Model

To build the 4-layer LSTM model, we used Sequential from Tensorflow.Keras. Four layers are as follows. One Embedding layer for embedding of the top 2000 words into a 64-dimensional embedding. One Bidirectional layer for the LSTM layer in order to have one output as the result of sequences of words as inputs. One Dense layer with Rectified Linear Unit (ReLU) as an activation function for this model. And finally, another Dense layer with the activation function of Sig-moid for defining that this model has the binary output. 0 for non-malicious and 1 as malicious data entry. The LSTM model was compiled with 'bina-ry_crossentropy' as the loss function and 'adam' as the optimizer algorithm in preparation to reduce the losses. The number of epochs for training the model was 10.

3.6   CNN Model

The CNN Sequential model has 6 different layers. One Embedding layer for the same reason as discussed earlier in the LSTM model. One Conv1D layer as our CNN layer with the 'same' padding function and 'ReLU' activation function. The third layer is the max-pooling layer. Then in the fourth layer, we flattened the inputs. Next, one Dense layer with the 'ReLU' activation function was added to the model. Finally, for the last layer, another Dense layer with the 'sigmoid' activation function was added to the model due to the fact that the binary classification is needed. The number of epochs for training the CNN model was 2.

4   Dashboard

To have an overview of all identified malicious records as a result of the model, a dashboard was designed and implemented. Consequently, a MySQL database with 4 different tables on a LightSail machine was created. The description of the tables is in this fashion.

- One table for saving top 10 suspicious IP addresses with the number of log entries they appeared.
- Another table for keeping top 10 enumeration attack event names with the number of appearances.
- Next table for preserving 24 h a day and the number of suspicious activities per hour.
- Last table for saving top 10 suspicious roles and their corresponding appearances in the dataset.

Results of the dashboard for the aforesaid dataset is illustrated in Fig. 2.

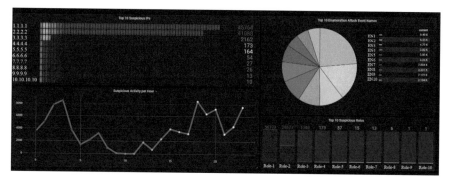

Fig. 2 Dashboard output

5   Resulting & Discussion

This section illustrates the results of the implemented models in categorizing the AWS dataset. The evaluation metrics which have been studied for the models are accuracy, loss, val_accuracy, val_loss, and their running time for each epoch.

5.1   LSTM Results

The LSTM model had remarkable results on the AWS dataset and performance metrics for the model are 99.94% for accuracy, 0.0030 for loss, 0.0024 for val_loss, and 99.96% for val_accuracy. The LSTM model fitted for 10 epochs that you can see the results in Figs. 3 and 4. Running time for each epoch in this model was 80 s.

5.2   CNN Results

The results for CNN model on AWS dataset were in this fashion. Accuracy of 99.94%, loss value of 0.0029, val_loss value of 0.0026, and val_accuracy 99.96% which indicate that the performance of the two models on AWS dataset were the same. It is noteworthy to say that the running time for each epoch in this model significantly decreased to just 8 s. Results details demonstrated in Figs. 5 and 6.

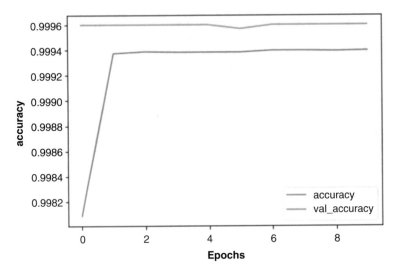

Fig. 3 LSTM accuracy

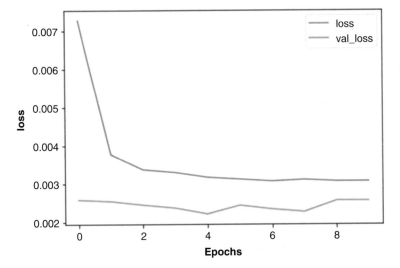

Fig. 4 LSTM loss

6 Conclusion & Future Work

Both LSTM and CNN models demonstrated high performance in anomaly detection on the AWS dataset. The validation accuracy for the two models was the same and was 99.96%. The validation loss for the LSTM model was better than the CNN model with a negligible difference of 0.0002 and was 0.0024. Furthermore,

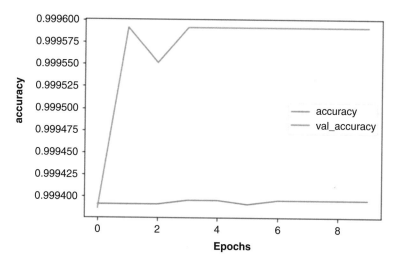

Fig. 5 CNN accuracy

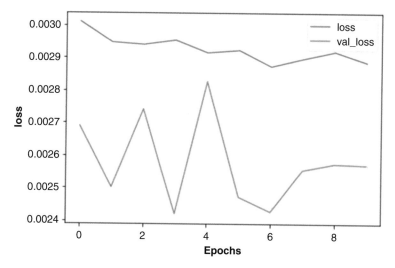

Fig. 6 CNN loss

the running time for each epoch in the CNN model was by far smaller than the LSTM model and was 8 s instead of 80 s. Preprocessing the dataset played the most important role in improving the performance of the models. Moreover, the visualization part of this project had done by showing the most important features, which show that threats in the logs such as IP addresses, roles, and even event names are mostly used for enumeration attacks. Future work should consider detecting enumeration attacks in other cloud environments such as Google Cloud Platform (GCP) and Microsoft Azure.

References

1. J. Baldwin, O.M.K. Alhawi, S. Shaughnessy, A. Akinbi, A. Dehghantanha, Emerging from the cloud: A bibliometric analysis of cloud forensics studies, in *Cyber Threat Intelligence*, (Springer, Cham, 2018), pp. 311–331
2. B. Blakeley, C. Cooney, A. Dehghantanha, R. Aspin, Cloud storage forensic: hubiC as a case-study, in *2015 IEEE 7th International Conference on Cloud Computing Technology and Science (CloudCom)*, (2015), pp. 536–541
3. Y. Teing, A. Dehghantanha, K.R. Choo, CloudMe forensics: A case of big data forensic investigation. Concurr. Comput. Pract. Exp. **30**(5), e4277 (2018)
4. L.S. Thiam, T. Dargahi, A. Dehghantanha, Bibliometric analysis on the rise of cloud security, in *Handbook of Big Data and IoT Security*, (Springer, Cham, 2019), pp. 329–344
5. A. Yazdinejad, R.M. Parizi, A. Dehghantanha, K.-K.R. Choo, P4-to-blockchain: A secure blockchain-enabled packet parser for software defined networking. Comput. Secur. **88** (2020). https://doi.org/10.1016/j.cose.2019.101629
6. S. Rahalkar, *Network Vulnerability Assessment: Identify Security Loopholes in Your Network's Infrastructure* (Packt Publishing Ltd, Birmingham, 2018)
7. A. Yazdinejad, R.M. Parizi, A. Bohlooli, A. Dehghantanha, K.-K.R. Choo, A high-performance framework for a network programmable packet processor using P4 and FPGA. J. Netw. Comput. Appl. **156**, 102564 (2020)
8. Q. Chen, G. Srivastava, R.M. Parizi, M. Aloqaily, I. Al Ridhawi, An incentive-aware blockchain-based solution for internet of fake media things. Inf. Process. Manag., 102370 (2020). https://doi.org/10.1016/j.ipm.2020.102370
9. A. Yazdinejad, R.M. Parizi, G. Srivastava, A. Dehghantanha, K.-K.R. Choo, Energy efficient decentralized authentication in internet of underwater things using blockchain, in *2019 IEEE Globecom Workshops (GC Wkshps)*, (2019), pp. 1–6
10. F.M.P.D. Johnson, *Robust Identity and Access Management for Cloud Systems* (2020). https://doi.org/10.7939/r3-ztwg-xm63
11. A. Yazdinejad, H. HaddadPajouh, A. Dehghantanha, R.M. Parizi, G. Srivastava, M.-Y. Chen, Cryptocurrency malware hunting: A deep recurrent neural network approach. Appl. Soft Comput. Elsevier **96**, 106630 (2020)
12. M. Aledhari, R. Razzak, R.M. Parizi, F. Saeed, Federated learning: A survey on enabling technologies, protocols, and applications. IEEE Access **8**, 140699–140725 (2020). https://doi.org/10.1109/ACCESS.2020.3013541
13. A. Yazdinejad, R.M. Parizi, A. Dehghantanha, H. Karimipour, G. Srivastava, M. Aledhari, Enabling drones in the internet of things with decentralized blockchain-based security. IEEE Internet Things J., 1 (2020). https://doi.org/10.1109/jiot.2020.3015382
14. A. Yazdinejad, R.M. Parizi, A. Dehghantanha, G. Srivastava, S. Mohan, A.M. Rababah, Cost optimization of secure routing with untrusted devices in software defined networking. J. Parallel Distrib. Comput. **143**, 36–46 (2020)
15. A. Zomaya et al., Cloud log forensics: Foundations, state of the art, and future directions. ACM Comput. Surv. **49**(1), 7 (2016)
16. P.N. Bahrami, A. Dehghantanha, T. Dargahi, R.M. Parizi, K.-K.R. Choo, H.H.S. Javadi, Cyber kill chain-based taxonomy of advanced persistent threat actors: Analogy of tactics, techniques, and procedures. J. Inf. Process. Syst. **15**(4), 865–889 (2019)
17. A. Azmoodeh, A. Dehghantanha, K.-K.R. Choo, Robust malware detection for internet of (battlefield) things devices using deep eigenspace learning. IEEE Trans. Sustain. Comput. **4**(1), 88–95 (2018)
18. S. Selvin, R. Vinayakumar, E.A. Gopalakrishnan,... Google Scholar. https://scholar.google.ca/scholar?hl=en&as_sdt=0%2C5&q=Selvin%2C+S.%2C+Vinayakumar%2C+R.%2C+Gopalakrishnan%2C+E.+A.%2C+Menon%2C+V.+K.%2C+%26+Soman%2C+K.+P.+%282017%2C+September%2. p. 282017

19. C. Zhou, C. Sun, Z. Liu, F. Lau, A C-LSTM neural network for text classification. arXiv Prepr. arXiv1511.08630 (2015)
20. M. Saharkhizan, A. Azmoodeh, A. Dehghantanha, K.-K.R. Choo, R.M. Parizi, An ensemble of deep recurrent neural networks for detecting IoT cyber attacks using network traffic. IEEE Internet Things J. **7**(9), 8852–8859 (2020). https://doi.org/10.1109/jiot.2020.2996425
21. H. HaddadPajouh, A. Dehghantanha, R. Khayami, K.-K.R. Choo, A deep recurrent neural network based approach for internet of things malware threat hunting. Futur. Gener. Comput. Syst. **85**, 88–96 (2018). https://doi.org/10.1016/j.future.2018.03.007
22. A. Graves, J. Schmidhuber, Framewise phoneme classification with bidirectional LSTM and other neural network architectures. Neural Netw. **18**(5–6), 602–610 (2005)
23. A. Yazdinejad, A. Bohlooli, K. Jamshidi, Efficient design and hardware implementation of the OpenFlow v1.3 Switch on the Virtex-6 FPGA ML605. J. Supercomput. **74**(3) (2018). https://doi.org/10.1007/s11227-017-2175-7
24. H. Kim, J. Kim, Y. Kim, I. Kim, K.J. Kim, Design of network threat detection and classification based on machine learning on cloud computing. Clust. Comput. **22**(1), 2341–2350 (2019)
25. E.K. Subramanian, L. Tamilselvan, A focus on future cloud: Machine learning-based cloud security. Serv. Oriented Comput. Appl. **13**(3), 237–249 (2019)
26. T. Salman, D. Bhamare, A. Erbad, R. Jain, M. Samaka, Machine learning for anomaly detection and categorization in multi-cloud environments, in *2017 IEEE 4th International Conference on Cyber Security and Cloud Computing (CSCloud)*, (2017), pp. 97–103
27. B. Gulmezoglu, T. Eisenbarth, B. Sunar, Cache-based application detection in the cloud using machine learning, in *Proceedings of the 2017 ACM on Asia Conference on Computer and Communications Security*, (2017), pp. 288–300
28. S. Rathore, P.K. Sharma, J.H. Park, XSSClassifier: An efficient XSS attack detection approach based on machine learning classifier on SNSs. J. Inf. Process. Syst. **13**(4), 1014–1028 (2017)
29. K.K. Nguyen, D.T. Hoang, D. Niyato, P. Wang, D. Nguyen, E. Dutkiewicz, Cyberattack detection in mobile cloud computing: A deep learning approach, in *2018 IEEE Wireless Communications and Networking Conference (WCNC)*, (2018), pp. 1–6
30. T. Schindler, Anomaly detection in log data using graph databases and machine learning to defend advanced persistent threats. arXiv Prepr. arXiv1802.00259 (2018)
31. W. Fang, X. Tan, D. Wilbur, Application of intrusion detection technology in network safety based on machine learning. Saf. Sci. **124**, 104604 (2020)
32. G. Yan, Q. Li, D. Guo, B. Li, AULD: Large scale suspicious DNS activities detection via unsupervised learning in advanced persistent threats. Sensors **19**(14), 3180 (2019)
33. K. Reghunath, Real-time intrusion detection system for big data. Int. J. Peer Peer Netw. (IJP2P) **8**(1) (2017). https://doi.org/10.5121/ijp2p.2017.8101
34. A. Azmoodeh, A. Dehghantanha, M. Conti, K.-K.R. Choo, Detecting crypto-ransomware in IoT networks based on energy consumption footprint. J. Ambient. Intell. Humaniz. Comput. **9**(4), 1141–1152 (2018)
35. G. Yan, Q. Li, D. Guo, X. Meng, Discovering suspicious APT behaviors by analyzing DNS activities. Sensors **20**(3), 731 (2020)
36. M.B. Seyyar, F.Ö. Çatak, E. Gül, Detection of attack-targeted scans from the Apache HTTP Server access logs. Appl. Comput. Inform. **14**(1), 28–36 (2018)
37. M. Saharkhizan, A. Azmoodeh, H. HaddadPajouh, A. Dehghantanha, R.M. Parizi, G. Srivastava, A hybrid deep generative local metric learning method for intrusion detection, in *Handbook of Big Data Privacy*, (Springer, 2020), pp. 343–357. https://doi.org/10.1007/978-3-030-38557-6_16
38. M. Zekri, S. El Kafhali, N. Aboutabit, Y. Saadi, DDoS attack detection using machine learning techniques in cloud computing environments, in *2017 3rd International Conference of Cloud Computing Technologies and Applications (CloudTech)*, (2017), pp. 1–7
39. M. Du, F. Li, G. Zheng, V. Srikumar, DeepLog, in *Proceedings of the 2017 ACM SIGSAC Conference on Computer and Communications Security-CCS*, vol. 17, (2017), pp. 1285–1298. https://doi.org/10.1145/3133956.3134015

40. S. Naseer et al., Enhanced network anomaly detection based on deep neural networks. IEEE Access **6**, 48231–48246 (2018)
41. A. Brown, A. Tuor, B. Hutchinson, N. Nichols, Recurrent neural network attention mechanisms for interpretable system log anomaly detection, in *Proceedings of the First Workshop on Machine Learning for Computing Systems*, (2018), pp. 1–8
42. M. Haselmann, D.P. Gruber, P. Tabatabai, Anomaly detection using deep learning based image completion, in *2018 17th IEEE International Conference on Machine Learning and Applications (ICMLA)*, (2018), pp. 1237–1242
43. cloud-service-enum/aws_service_enum at master · NotSoSecure/cloud-service-enum · GitHub. https://github.com/NotSoSecure/cloud-service-enum/tree/master/aws_service_enum. Accessed 16 Sep 2020
44. World Health Organization, et al., GitHub – toniblyx/my-arsenal-of-aws-security-tools: List of open source tools for AWS security: defensive, offensive, auditing, DFIR, etc. https://github.com/toniblyx/my-arsenal-of-aws-security-tools. Accessed 16 Sep 2020. Osteoarthr. Cartil. **28**(2), 1–43. https://doi.org/10.18420/in2017
45. eSentire | Modern threat hunting for the digital age | eSentire. https://www.esentire.com/. Accessed 16 Sep 2020

Cyber Threat Attribution with Multi-View Heuristic Analysis

Dilip Sahoo

1 Introduction

Past decades has witnessed a considerable rise for using digitalized system for different aspects of our modern life ranging from agriculture [1] and health [2] to power grid [3] and transportation [4] which has motivated cyber attackers for designing sophisticated and target -specific attacks against such infrastructures [5–8]. In the context of cybersecurity, threat attribution is a fundamental step to find out who is behind an attack. Ascribing a group or agency to a threat helps the security professionals to take appropriate countermeasures to protect the individuals and organizations. APTs are the most challenging which are on the rise for security professionals to defend against [5, 9, 10]. The APT groups use specific TTP to target, penetrate, and exploit organizations. Because of the sophisticated nature of the attack strategies adopted by the APT actors, it is not easy to attribute them against an attack [11]. A report from McAfee claims that most APT attacks are interrelated in their nature of the attack and they have similar target organizations [12]. The general characteristics of APTs are that they are sophisticated, targeted, and evasive, the attack adopts to security measures, and has multiple attack vectors.

The APT attacks are evasive that use several methods to stay undetected like using commonly accepted protocols for sending threat contents. They use custom encryption techniques to overcome firewall detection. It is common to notice various detection evading and code obfuscation techniques used in many malware variants. These evasive techniques create confusion and may sidetrack the analysts while focusing on a specific characteristic (from a single-view) of the malware.

D. Sahoo (✉)
Cyber Science Lab, University of Guelph, Guelph, ON, Canada
e-mail: dilip@cybersciencelab.org

© Springer Nature Switzerland AG 2021
K.-K. R. Choo, A. Dehghantanha (eds.), *Handbook of Big Data Analytics and Forensics*, https://doi.org/10.1007/978-3-030-74753-4_4

In most APT campaigns, it is noticed they launch very sophisticated attacks and target a particular type of organization. For example, the Stuxnet campaign was targeting centrifuges that use programmable logic controllers (PLCs) manufactured by Siemens. Stuxnet was using an extremely sophisticated worm that exploits zero-day vulnerabilities of windows systems [4]. These types of attacks can easily drop and install payloads in the target system as the attacks are designed exclusively for the target systems. It is often impossible to detect such attacks by traditional Antivirus or Intrusion Detection Systems. In the last decade, behavioral analysis of malware files in a sandbox environment has become popular. In this method, the researchers observe malware behaviors like network traffic, system calls, registry updates, etc. at runtime by executing the malware in an isolated environment. This method is very effective against the unknown malware payloads that belong to APTs. However, this method is a time-consuming process and not useful in realtime detection scenarios. Also, sometimes the sophisticated malware programs can distinguish the sandbox environment from a real environment and behave differently.

On the other hand heuristic analysis, uses ML algorithms to train the systems with malware behavior. Such systems can be trained on the existing known malware file features like Opcode, Bytecode, Header details. The features can be fed to different Machine Learning [13–15] and Deep Learning [16–19] classifiers to perform the classification task. Heuristic analysis is effective against both unknown and metamorphic malware detection. A major pitfall of machine learning classifiers is that the output can get biased based on the training data. Sometimes training data may contain biases that can result in improper outcomes and impact the overall performance of the ML classifiers [20] To address the issue of biased prediction by a classifier, it is important to feed the ML classifier with balanced data [21].

In the experiment, research is conducted on more than 3000 malware files from 12 APT families. A multi-view approach similar to [22, 23] is followed for the experiment. Below are the major research contributions made as part of this work.

I. Eleven different views were created based on the extracted Opcode, Bytecode, and Header features to look at the files under observation from different aspects. This helps to make the system resilient against obfuscation and evasion techniques.

II. Five different machine learning algorithms namely SVM, Decision Tree, KNN, MLP, and Fair Clustering were evaluated with all the views.

III. A SMOTE data set was developed with balanced distributions of data samples to reduce bias in favor of any particular class.

IV. A Multi-View prediction approach was adopted by combining the individual predictions from the single-views based on majority class prediction and accuracy(%).

Section 2 of this paper contains details on related work done recently for APT threat detection. Section 3 contains a brief description of the dataset used in the experiment. Section 4 details our experiment methodology which is followed by our Experimentation and Results in Sect. 5. Section 6 highlights a comparison of our findings with related works. Section 7 presents Our concluding statements and avenue for future work.

2   Related Work

Knowledge of the threat source increases the confidence of the security professionals during incident triaging and later with the incident response phase. It also helps them to decide the next course of action in a time-efficient manner due to the additional supplements of information regarding the TTP used by the attacker groups. Due to the substantial benefits of threat source attribution, various approaches have been taken by researchers to effectively automate the process of cyber threat attribution [24–31].

The paper [32] combines several individually contributed papers and provides a basic understanding of APTs along with explanations, examples, case studies on the APT phenomenon, their characteristics, APT attack stages, and how they should be handled. The paper discusses APT definitions considering the viewpoints and case study reports provided by leading security organizations and government agencies. The papers suggest that any organization should consider the APT threat seriously due to the targeted nature and suggests different ways to efficiently protect against APT campaign attacks.

Several types of research have been conducted to detect and prevent APT attacks. Implementation of traffic data analysis is one of the most popular approaches suggested by several researchers [33–35]. Trafic data analysis is conducted by analysis of network protocols, carried operations, and data that flows through the network. Researchers in [36] suggested a combination of traffic data analysis with an open-source intrusion detection system that analyzes the protocols used, requests sent, and uses filtering using black-list.

Pattern recognition is another popular approach to detect and prevent and APT attacks. In this approach, malicious programs are considered to be similar and they are distinguished from the benign applications by tracing their operational similarities and differences. Authors in [36] suggest a single layer pattern recognition approach. It is also common that several methods are combined to create a system that can protect against the APT threat than the individual methods. Moon et al. [37] and Vert et al. [38] used a combination of pattern recognition and multilayer security for detection and protection against APT threats at different security layers.

Heuristic analysis using ML classifiers is becoming more popular than traditional detection methods in the last decade. Authors in [39] present an interesting approach to detect malware Application. The proposed system uses the Application Programming Interfaces called by the malware program and technical PE features to classify malware files. It uses the chi-square (KHI^2) measure and Phi (φ) coefficient for considering features by relevance. The system could accomplish binary classification with 98% accuracy in a time-efficient manner.

Authors in [22] implemented a Multi-View Fuzzy Consensus Clustering Model for Malware Threat Attribution. The suggested approach uses 12 views to attribute the malware from five APT classes. It implements a fuzzy pattern tree, multi-modal fuzzy classifier, and consensus clustering technique to analyze the malware behavior. The suggested system could perform threat attribution with 95% accuracy.

3   Dataset

A dataset of 3594 malware file samples [40] belonging to 12 different APT groups namely APT1, APT10, APT19, APT21, APT28, APT29, APT30, DarkHotel, EnergeticBear, EquationGroup, GorgonGroup, and Winnti was used for the experiment. These APTs were alleged to be sponsored by five different nation-states. Each of the malware files was processed using additional python scripts to extract details of the Opcode, Bytecode, and Header information and to create multiple views based on that information. Details of the data processing and view creation are discussed in later Sects. 4.1 and 4.2. Table 1 illustrates the number of malware samples collected against each APT group.

4   Methodology

In this section, the detailed steps taken during the experiment to implement a multi-view-based malware attribution model are described. The Malware files for 12 different APT groups were collected and processed to create the multi-view data samples. Later these multi-view data samples were used to attribute each malware

Table 1 APT-Malware data description

Sl no	APT group name	No of malware files	Nation-state
1	APT1	405	China
2	APT10	244	China
3	APT19	32	China
4	APT21	106	China
5	APT28	214	Russia
6	APT29	281	Russia
7	APT30	164	China
8	DarkHotel	273	North Korea
9	EnergeticBear	132	Russia
10	EquationGroup	395	USA
11	GorgonGroup	961	Pakistan
12	Winnti	387	China

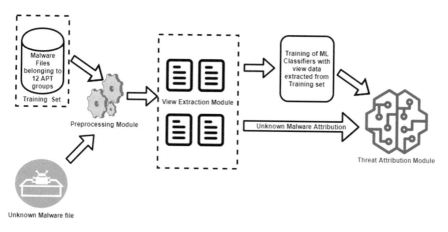

Fig. 1 The Multi-view malware attribution system

file to an APT group. During the experiment, five Machine learning classifiers were trained and evaluated in terms of accuracy. Finally, the best detection models for each view were identified and implemented for malware attribution. The proposed system consists of three important modules namely the pre-processing module, the view extraction module, and a threat attribution module. Figure 1 illustrates the modules and steps involved in our threat attribution experiment.

4.1 Preprocessing and View Extraction

The raw malicious files were processed using custom python scripts to extract information on details of the Opcode, Bytecode, and Header from each of them. This extracted information is treated as the foundation of our multi-view approach to perform further heuristic analysis using ML classifiers. The processing and creation of each view are described in the below Sects. 4.1.1, 4.1.2, and 4.1.3.

4.1.1 Opcode

Opcodes are the assembly instructions present in the malware executable files. To extract the opcode information from the malware executable files, the Linux 'Objdump' command was used to disassemble the binary files. Then a dictionary of all available unique opcodes was created which became the base for further processing of Opcode based views that is referred to as Opcode_Dic. Five different types of view samples were derived from the Opcode data extracted from the Malware binary files namely Binary, Count, Frequency, Term Frequency-Inverse Document Frequency (TFIDF), and Eigen Vector. For the creation of Binary, Count,

Frequency, and TFIDF views, the Text tokenization utility class using Keras [41] was used. The Eigen Vector view was created following the method proposed by Hashem et al. [42]. ML classifiers were trained with all the Opcode based views and their performance in terms of accuracy was noted against each view. This gave us a holistic comprehension of how different ML classifiers perform for different views. It is worth noting that all the Opcode based views were derived from the same Opcode information extracted from the malware files and by using different processing techniques.

Binary

This view is named as binary due to the nature of the data values present inside this view which is either '0' or '1'.The Opcode Binary view sample was created by checking whether a particular Opcode value of the Opcode_Dic is present in a malware binary file or not. A value of '1' was assigned if it is present and '0' if it is not. The Opcode_Dic is considered as the base document to be referred for creating this view. Hence, the number of columns or features for this view remains constant for all the malware files. The number of columns is the same as the unique number of Opcodes present in the Opcode_Dic.

Count

In this view, the data values are represented as the count of each Opcode of the Opcode_Dic file, that is present in malware files. Hence, in contrast to the binary view, this view represents the actual count value of the Opcodes instead of '0' and '1'. The number of columns is the same as the unique number of Opcodes present in the Opcode_Dic.

Frequency

This view represents the frequency value of each Opcode from Opcode_Dic file as a ratio of the Opcodes present in the malware files. The number of columns is the same as the unique number of Opcodes present in the Opcode_Dic.

TFIDF

The TFIDF view represents the Term Frequency-Inverse Document Frequency score of each Opcode present in the malware files. The number of columns is the same as the unique number of Opcodes present in the Opcode_Dic.

Eigen Vector

The eigenvector view uses function call graph as the signature of the program proposed by authors in [43]. It uses the graph representation of the program and apply the mathematical equation to detect malware.

4.1.2 Bytecode

The Bytecode sequences of the malware files were extracted using custom python scripts. The bytecode values lie between 0 and 255. Hence, the bytecode dictionary file was created with all the values ranging from 0 to 255 and referred to as Bytecode_Dic. Five different sample views Binary, Count, Frequency, TFIDF, and Eigen Vector were created from the extracted bytecode values like the Opcode views discussed in Sect. 4.1.1. Finally, the ML classifiers were evaluated for each of the Bytecode views.

4.1.3 Header

The header view represents the header information gathered from the malware Portable Executable (PE) files. The header fields were extracted from the PE file header and PE optional header sections using python libraries like: *'pefile'* [44] *and 'lief'* [45]. The field information like 'Machine', 'SizeOfOptionalHeader', and 'Characteristics' was extracted from the PE file header section of the malware PE files. Similarly, from PE Optional header section, the fields: 'MajorLinkerVersion', 'MinorLinkerVersion', 'SizeOfCode' etc. were extracted. Figure 2 shows the PE file header and PE Optional header information of a sample malware PE file. Due to the huge variance between the raw data collected from the header section, the raw data was later normalized using a logarithmic function. The normalized data were used to create the final Header view.

```
Header
------
Signature:                    50 45 0 0
Machine:                      I386
Number Of Sections:           4
Pointer To Symbol Table:      0
Number Of Symbols:            0
Size Of Optional Header:      e0
Characteristics:              RELOCS_STRIPPED - EXECUTABLE_IMAGE - LINE_NUMS_STRIPPED - LOCAL_SYMS_STRIPPED - CHARA_32BIT_MAC
HINE
Time Date Stamp:              4e408f53

Optional Header
---------------
Magic:                        10b
Major Linker Version:         6
Minor Linker Version:         0
Size Of Code:                 1e00
Size Of Initialized Data:     1600
Size Of Uninitialized Data:   0
```

Fig. 2 File header and Optional Header section fields of a malware PE file

4.2 Data Balancing Using Synthetic Minority Over-Sampling Technique (SMOTE)

After the creation of the views, it was observed that the data samples inside the views were not balanced. This is because certain APT groups had more malware samples than others. For example, there were only 32 malware samples belong to APT19 (shown in Table 1) which is considerably less compared to other APT groups. An imbalanced dataset can cause biased results and poor predictive performance, especially for the minority class. Hence, the imbalance dataset poses a challenge to the overall ML algorithm performance [21, 46]. To overcome the issues of the imbalanced data in the views, the SMOTE technique was used to balance the dataset by upsampling the minority class data. The SMOTE enhanced views were then used to train the ML classifiers.

4.3 Machine Learning Classifier Phase

In this phase, four well-known ML classifiers namely SVM, Decision Tree, KNN, and MLP were implemented from the open-source scikit-learn library (https://scikit-learn.org). Also, a Fair Clustering algorithm suggested by Backurs et al. [47] was adopted for the experiment. Each of the above-mentioned classifiers was evaluated with all the view samples' data explained in Sect. 4.1. The experiments were conducted in a 13 GB RAM Windows 10 virtual machine with 2.21 GHz 64-bit intel i7 processor. Another 4 GB Ubuntu 20.04 virtual machine was used for the extraction of Opcode information from the malware files. Python 3.6.5 and MATLAB engine were used with jupyter notebook.

4.3.1 Support Vector Machine (SVM)

SVM is a simple algorithm that produces significant accuracy with less computational power [48]. The SVM algorithm finds hyperplane to classify N-dimensional data where 'N' is the number of features in the dataset. Due to the multiclass nature of the sample views' data, 'decision_function_shape' parameter value as 'one-vs-one (ovo)' was used which is a common approach followed during multi-class classification.

4.3.2 Decision Tree

A decision tree classifier can be used for both classification and regression tasks [49]. The feature importance and relations can be visualized clearly in a decision tree. It uses a greedy algorithm to lower costs.

4.3.3   K-Nearest Neighbour (KNN)

The KNN is an unsupervised ML algorithm that predicts the label of a new point from the testing sample by checking the label of 'K' predefined training samples which are close in distance. The value of 'K' was kept as K = 5 during the experiment which is the default value and gave us an optimum result.

4.3.4   Multi-layer Perceptron (MLP)

MLP is a type of neural network which is a deep-learning-based classifier. MLP is powerful because of multiple layers but can be a computationally expensive classifier. During the experiment, three hidden layers with 100 nodes each were used and the maximum number of iteration was set as 200 epochs.

4.3.5   Fair Clustering

A fair clustering approach was suggested by Backurs et al. [47] that provides fairness as well as scalability to the clustering algorithm and runs in near-linear time. Because of the additional benefits, it was decided to implement this approach instead of the traditional k-median clustering algorithm. The elbow method was used to find an optimum cluster value 'k' which was set to k = 20 during the experiment.

5   Experiments and Results

This section describes the details of the experiment conducted and highlights the results. Section 5.1 describes the evaluation measures adopted for the assessment. The experiment was conducted in two phases, in the first phase, described in Sect. 5.2, the original view data obtained from the raw malware samples were used to train five Machine Learning classifiers (details of ML classifiers are mentioned in Sect. 4.3) and the results were analyzed. After analyzing the results obtained from the first phase, the second phase of the experiment was conducted by feeding SMOTE enhanced balanced datasets to the four ML classifiers that performed best during the first phase. The details of the second phase experiment and its results are highlighted in Sect. 5.3. Finally, Sect. 5.4 demonstrates the results obtained using a multi-view prediction approach.

5.1 Evaluation measures

The threat attribution model in the experiment is a multi-class classification model where each threat actor is considered as a class. For instance, considering the confusion matrix we have obtained during our experiment from the OPCODE_TFIDF view illustrated in Fig. 3. There are 12 different classes denoting 12 APT groups. The multiclass classification model matrices can be understood as a set of several binary class classification models (where there are only two classes as 'Positive' or 'Negative'). For example, in our classifier, if we consider a malware file belong class 'APT1' then a True Positive occurs, when the malware is correctly predicted to be of class 'APT1'. Any other prediction will be considered to be a 'false negative'. In the multi-class classification, the positive and negative will depend on the true label of a sample and can change based on the object label. It means that for a given prediction, there will be multiple classes as 'true negative'. For instance, while considering class APT10, if a malware file that originally belongs to APT21 is predicted to be of any class (i.e. APT1, APT19, APT21, . . . , Winnti) other than APT10, then it will be considered as true negative for the class APT10.

The evaluation measures for the experiment are derived from the confusion matrix. A common Confusion matrix represents the summary of all the predicted results of a classifier in terms of the number of True Positives (TP), True Negatives (TN), False Positives (FP) and False Negatives (FN). The diagonal elements in the confusion metrics represent the correctly classified samples for each APT. 'Accuracy' is the number of samples correctly identified as true positive or true

	APT1	APT10	APT19	APT21	APT28	APT29	APT30	DarkHotel	EnergeticBear	EquationGroup	GorgonGroup	Winnti
APT1	372	2	3	1	1	4	3	9	0	0	2	3
APT10	1	381	1	1	3	2	0	7	0	0	4	0
APT19	0	0	400	0	0	0	0	0	0	0	0	0
APT21	0	0	0	399	1	0	0	0	0	0	0	0
APT28	0	0	0	0	392	2	1	2	0	0	2	1
APT29	0	1	1	0	3	384	0	1	0	0	4	6
APT30	1	0	0	0	0	0	396	1	0	0	1	1
DarkHotel	1	7	0	1	1	5	5	371	0	0	4	5
EnergeticBear	0	0	0	0	1	0	0	0	399	0	0	0
EquationGroup	0	0	0	0	0	0	1	0	0	399	0	0
GorgonGroup	1	1	2	0	1	4	2	4	1	0	374	10
Winnti	4	1	2	0	2	3	0	3	0	1	8	376

Fig. 3 OPCODE_TFIDF Confusion Matrix using MLP

negative out of all items. 'Precision' is the number of correctly identified positive samples out of all the positive predictions. 'Recall' also known as 'True positive rate' or 'Sensitivity rate' is the number of correctly predicted positive samples out of all the actual positives. F1-Score is the harmonic average of precision and recall and determines the effectiveness of the identification.

TP: An APT actor with a true label as positive predicted correctly to be positive
TN: An APT actor with a true label as negative predicted correctly to be negative
FP: An APT actor with a true label as negative predicted incorrectly to be positive
FN: An APT actor with a true label as positive predicted incorrectly to be negative

$$Accuracy = \frac{TP + TN}{TP + TN + FP + FN}$$

$$Precision = \frac{TP}{TP + FP}$$

$$Recall = \frac{TP}{TP + FN}$$

$$F1 - score = 2 * \left(\frac{Precision * Recall}{Precision + Recall} \right)$$

5.1.1 Single-View Prediction vs Multi-View Prediction

The classifiers (described in Sect. 4.3) evaluated with the individual views extracted during the preprocessing phase (described in Sect. 4.1) referred to as 'Single-View prediction'. Single-View predictions include assessments from Opcode, Bytecode (binary, count, frequency, and TFIDF) and header views. Optimization of the prediction results was done by leveraging multiple Single-View predictions. The high-level approach is to consider the prediction from the majority of the individual Single-View as the final Multi-View outcome. If there the majority APT actor cannot be decided between the Single-Views (When every Single-View predicted a different APT Class) then weightage is given to individual Single-View predictions based on the accuracy(%). The final Multi-View predictions will be determined by combining the individual predictions from the Single-View predictions. Figures 4 and 5 illustrates scenarios of Multi-View prediction.

Multi-View Prediction
Scenario-1

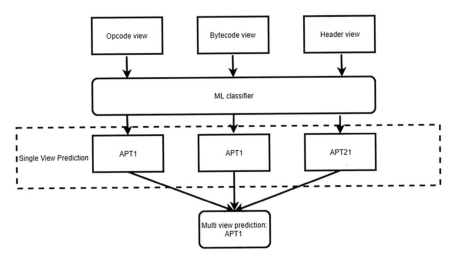

Fig. 4 Multi-View Prediction using majority class predicted by individual Single-View

Scenario-2

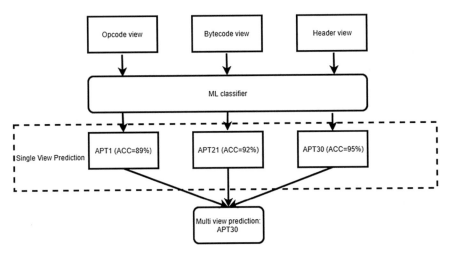

Fig. 5 Multi-View Prediction using highest accuracy class predicted by individual Single-View

5.2 Experiment Phase-1 and Results

In this phase of the experiment, the individual Single-Views that belong to Opcode, Bytecode (binary, count, frequency, tfidf, Eigen Vector), and Header are evaluated with the classifiers individually. The data samples from original views were used to evaluate the classifiers at this phase using a tenfold cross-validation technique. It was observed that OPCODE_TFIDF view and BYTECODE_FREQUENCY views gave the best results among other Opcode and Bytecode based views respectively. It was noticed that SVM, DT, KNN, and MLP classifiers outperformed the FAIR_CLUSTERING classifier in terms of accuracy. The overall accuracy results obtained using tenfold cross-validation from the experiment phase-1 are summarized in Table 2.

5.3 Experiment Phase-2 and Results

After analyzing the overall outcome of the experiment phase-1, the views that provided the best accuracy under each category (i.e. Opcode, Bytecode, and Header) are selected for the experiment phase-2. To further optimize the performance of the classifiers, the data balancing of the selected views was done using SMOTE technique described in Sect. 4.2. Finally, each view is evaluated with the top 4 classifiers that performed best in experiment phase-1. In this phase, a 5–10% improvement in accuracy was observed for each view for the four classifiers under consideration. Table 3 illustrates the summary of accuracy improvement in experiment phase-2.

Table 2 Summary of accuracy(%) from experiment phase-1

View name	Accuracy (%)				
	SVM	DT	KNN	MLP	FAIR_CLUSTERING
OPCODE_BINARY	77	85	82	89	49
OPCODE_COUNT	53	89	81	86	64
OPCODE_FREQUENCY	75	88	82	89	58
OPCODE_TFIDF	73	89	85	91	57
OPCODE_EIGEN_VECTOR	51	67	52	68	28
BYTECODE_BINARY	30	31	22	31	24
BYTECODE_COUNT	58	82	81	81	47
BYTECODE_FREQUENCY	83	80	81	88	54
BYTECODE_TFIDF	72	82	82	88	49
BYTECODE_EIGEN_VECTOR	44	52	30	39	35
HEADER	82	89	86	91	62

Table 3 Summary of improvement of accuracy in experiment phase-2

View name	SVM accuracy (%)		DT accuracy (%)		KNN accuracy (%)		MLP accuracy (%)	
	Original	SMOTE enhanced	Original	SMOTE enhanced	Original	SMOTE enhanced	Original	SMOTE enhanced
OPCODE_TFIDF	73	83	89	92	85	92	91	97
BYTECODE_FREQUENCY	83	93	80	92	81	95	88	98
HEADER	82	87	89	93	86	93	91	96

Clearly, the MLP classifier prediction results were best among other classifiers that are evaluated during the experiments. MLP is a powerful deep learning algorithm and hence it is important to analyze performance matrices like overall runtime and evaluation measures of individual APT class. Table 4 illustrates a detailed APT class-wise prediction result for the views along with runtime and overall accuracy.

5.4 *Multi-View Prediction*

After having the best prediction results from individual single views during experiment phase-2, the multi-view prediction approach (as described in Sect. 5.1.1) was adopted to further optimize the prediction performance. Table 5 shows the prediction performance obtained with the multi-view approach using a subset of the original data and Fig. 6 shows the confusion matrix of the multi-view prediction. The multi-view prediction provided an accuracy of 99% which is higher than the individual single-views observed in experiment phase-2.

6 Results Comparison

This section, an efficiency evaluation of our system is done by comparing the results with some previously cited similar research work. Table 6 shows the comparison of results between our experiment and other similar work.

From the results shown in Table 6, It can be seen that the experiment provides higher accuracy than the other two presented work. The proposed system uses an MLP classifier which is a deep learning-based algorithm and is more complex than the classifiers used in other systems. Although the deep learning algorithms are complex and need higher runtime they provide higher accuracy.

The system proposed by Mohamed et al. provides 98% accuracy and takes only 0.090 s for the categorization process. However, his system only does binary classification. In contrast, our proposed system provides 99% accuracy for a multiclass classification for 12 different APT classes. The system by Hamed et al. provides an overall accuracy of 95% for a multiclass classification for 5 APT classes. The experiment by Hamed et al. uses a lesser number of APT classes then our proposed system. Hence, it can be seen that our proposed system performs better in terms of accuracy and is more efficient to perform the multiclass classification task.

Table 4 Detail results from MLP classifier for the views of Opcode, Bytecode and Header category that performed best during the experiment

View name	Classifier	APT Name	Precision	Recall	f1-score	Run time (Sec)	Overall accuracy (%)
OPCODE_ TFIDF	MLP	APT1	0.98	0.93	0.95	255.24	97
		APT10	0.97	0.95	0.96		
		APT19	0.98	1	0.99		
		APT21	0.99	1	1		
		APT28	0.97	0.98	0.97		
		APT29	0.95	0.96	0.96		
		APT30	0.97	0.99	0.98		
		DarkHotel	0.93	0.93	0.93		
		EnergeticBear	1	1	1		
		EquationGroup	1	1	1		
		GorgonGroup	0.94	0.94	0.94		
		Winnti	0.94	0.94	0.94		
BYTECODE_ FREQUENCY	MLP	APT1	0.98	0.98	0.98	312.94	98
		APT10	0.96	0.97	0.97		
		APT19	0.99	1	0.99		
		APT21	0.99	1	0.99		
		APT28	0.95	0.99	0.97		
		APT29	0.97	0.97	0.97		
		APT30	0.99	0.99	0.99		
		DarkHotel	0.97	0.98	0.97		
		EnergeticBear	0.99	1	1		
		EquationGroup	1	1	1		
		GorgonGroup	0.97	0.88	0.92		
		Winnti	0.96	0.96	0.96		
VG_ HEADER	MLP	APT1	0.93	0.92	0.92	160.1	96
		APT10	0.92	0.94	0.93		
		APT19	0.97	1	0.98		
		APT21	0.97	0.99	0.98		
		APT28	0.96	0.96	0.96		
		APT29	0.95	0.95	0.95		
		APT30	0.96	0.97	0.96		
		DarkHotel	0.92	0.9	0.91		
		EnergeticBear	0.99	0.98	0.99		
		EquationGroup	1	1	1		
		GorgonGroup	0.96	0.95	0.96		
		Winnti	0.95	0.93	0.94		

Table 5 Overall performance Matrix of Multi-view

View name	Classifier	APT name	Precision	Recall	f1-score	Overall accuracy (%)
Multi_View	MLP	APT1	1	1	1	99
		APT10	0.995	1	0.997	
		APT19	1	1	1	
		APT21	1	1	1	
		APT28	1	1	1	
		APT29	1	1	1	
		APT30	1	1	1	
		DarkHotel	1	0.996	0.998	
		EnergeticBear	1	1	1	
		EquationGroup	1	1	1	
		GorgonGroup	1	1	1	
		Winnti	1	1	1	

	APT1	APT10	APT19	APT21	APT28	APT29	APT30	DarkHotel	EnergeticBear	EquationGroup	GorgonGroup	Winnti
APT1	400	0	0	0	0	0	0	0	0	0	0	0
APT10	0	227	0	0	0	0	0	1	0	0	0	0
APT19	0	0	31	0	0	0	0	0	0	0	0	0
APT21	0	0	0	86	0	0	0	0	0	0	0	0
APT28	0	0	0	0	162	0	0	0	0	0	0	0
APT29	0	0	0	0	0	254	0	0	0	0	0	0
APT30	0	0	0	0	0	0	159	0	0	0	0	0
DarkHotel	0	0	0	0	0	0	0	268	0	0	0	0
EnergeticBear	0	0	0	0	0	0	0	0	130	0	0	0
EquationGroup	0	0	0	0	0	0	0	0	0	395	0	0
GorgonGroup	0	0	0	0	0	0	0	0	0	0	278	0
Winnti	0	0	0	0	0	0	0	0	0	0	0	384

Fig. 6 Multi-View Confusion matrix

Table 6 Result comparison from similar previously cited research work

Method	Classifier	Classification type	Accuracy
Our Method	MLP	Multi-Class (12 APT actor)	99%
Hamed et al.	Fuzzy Classifier	Multi-Class (5 APT actors)	95%
Mohamed et al.	BJ-48	Binary (Malware or Benign)	98%

7 Conclusion and Future Work

In this work, the malware file features are extracted, analyzed, and successfully attributed to their source APT actor with 99% accuracy. More than 3000 malware samples that belong to 12 APT groups were used in the experiment and 11 different views were created from the extracted features of Opcodes, Bytecodes, and Headers. Hence, the experiment approach deals with a comprehensive analysis during the attribution process that makes the system resilient towards the complex obfuscation and evasion techniques, commonly used during APT campaigns. The multi-view approach used for the threat attribution provides the final prediction by considering the underlined results from individual single-views. It could optimize the final prediction accuracy to 99% which is higher than the respective single-views.

Because of the complex nature of APT attacks, it is not always easy to attribute a threat vector to its source. However, heuristic analysis using Machine Learning algorithms can be used to automate the threat attribution process with higher accuracy. The threat attribution results can contribute significantly to improve the decision-making process and reduce time during an investigation.

During the experiment, malware data belong to 12 APT groups were used and five different ML classifiers were evaluated against it. It was observed that the performance of the ML classifiers varied with respect to different input views. SMOTE technique was used to balance the dataset. However, the SMOTE technique provided synthetic data samples that are different from the real data. Higher quality real-world data can help towards creating a more reliable system. More ML algorithms can be evaluated against the data sets to optimize the system.

Acknowledgments The author would like to thank Dr. Ali Dehghantanha and Hamed Haddad-pajouh for their valuable review during the research work. The author would also like to express gratitude to the creators of the malware database that was used in the experiment.

References

1. S. Nakhodchi, A. Dehghantanha, H. Karimipour, Privacy and security in smart and precision farming: A bibliometric analysis, in *Handbook of Big Data Privacy*, (Springer, Cham, 2020), pp. 305–318
2. A. Yazdinejad, G. Srivastava, R.M. Parizi, A. Dehghantanha, K.-K.R. Choo, M. Aledhari, Decentralized authentication of distributed patients in hospital networks using blockchain. IEEE J. Biomed. Heal. Inform. **24**(8), 2146–2156 (2020)
3. H.M. Rouzbahani, Z. Faraji, M. Amiri-Zarandi, H. Karimipour, AI-enabled security monitoring in smart cyber physical grids, in *Security of Cyber-Physical Systems*, (Springer, Cham, 2020), pp. 145–167. https://doi.org/10.1007/978-3-030-45541-5_8
4. A. Yazdinejad, R.M. Parizi, A. Dehghantanha, H. Karimipour, G. Srivastava, M. Aledhari, Enabling drones in the internet of things with decentralized blockchain-based security. IEEE Internet Things J., 1 (2020). https://doi.org/10.1109/jiot.2020.3015382
5. S. Grooby, T. Dargahi, A. Dehghantanha, Protecting IoT and ICS platforms against advanced persistent threat actors: Analysis of APT1, Silent Chollima and molerats, in *Handbook of Big Data and IoT Security*, (Springer, Cham, 2019), pp. 225–255

6. P.J. Taylor, T. Dargahi, A. Dehghantanha, Analysis of apt actors targeting IoT and big data systems: Shell_crew, nettraveler, projectsauron, copykittens, volatile cedar and transparent tribe as a case study, in *Handbook of Big Data and IoT Security*, (Springer, Cham, 2019), pp. 257–272
7. A. Yazdinejad, A. Bohlooli, K. Jamshidi, Efficient design and hardware implementation of the OpenFlow v1.3 Switch on the Virtex-6 FPGA ML605. J. Supercomput. **74**(3) (2018). https://doi.org/10.1007/s11227-017-2175-7
8. A. Yazdinejad, R.M. Parizi, A. Dehghantanha, K.-K.R. Choo, P4-to-blockchain: A secure blockchain-enabled packet parser for software defined networking. Comput. Secur. **88** (2020). https://doi.org/10.1016/j.cose.2019.101629
9. N. Pitropakis, E. Panaousis, A. Giannakoulias, G. Kalpakis, R.D. Rodriguez, P. Sarigiannidis, An enhanced cyber attack attribution framework, in *International Conference on Trust and Privacy in Digital Business*, (2018), pp. 213–228
10. P.N. Bahrami, A. Dehghantanha, T. Dargahi, R.M. Parizi, K.-K.R. Choo, H.H.S. Javadi, Cyber kill chain-based taxonomy of advanced persistent threat actors: Analogy of tactics, techniques, and procedures. J. Inf. Process. Syst. **15**(4), 865–889 (2019)
11. Advanced Persistent Threat Groups. FireEye, https://www.fireeye.com/current-threats/aptgroups.html. Accessed 5 July 2020
12. D. Alperovitch, Revealed: Operation Shady RAT, p. 14
13. A. Azmoodeh, A. Dehghantanha, M. Conti, K.-K.R. Choo, Detecting crypto-ransomware in IoT networks based on energy consumption footprint. J. Ambient. Intell. Humaniz. Comput. **9**(4), 1141–1152 (2018)
14. H.H. Pajouh, R. Javidan, R. Khayami, D. Ali, K.-K.R. Choo, A two-layer dimension reduction and two-tier classification model for anomaly-based intrusion detection in IoT backbone networks. IEEE Trans. Emerg. Top. Comput. **99**, 1 (2016)
15. E.M. Dovom, A. Azmoodeh, A. Dehghantanha, D.E. Newton, R.M. Parizi, H. Karimipour, Fuzzy pattern tree for edge malware detection and categorization in IoT. J. Syst. Archit. **97**, 1–7 (2019)
16. A. Azmoodeh, A. Dehghantanha, K.-K.R. Choo, Robust malware detection for internet of (battlefield) things devices using deep eigenspace learning. IEEE Trans. Sustain. Comput. **4**(1), 88–95 (2018)
17. M. Saharkhizan, A. Azmoodeh, A. Dehghantanha, K.-K.R. Choo, R.M. Parizi, An ensemble of deep recurrent neural networks for detecting IoT cyber attacks using network traffic. IEEE Internet Things J. **7**(9), 8852–8859 (2020). https://doi.org/10.1109/jiot.2020.2996425
18. M. Saharkhizan, A. Azmoodeh, H. HaddadPajouh, A. Dehghantanha, R.M. Parizi, G. Srivastava, A hybrid deep generative local metric learning method for intrusion detection, in *Handbook of Big Data Privacy*, (Springer, Cham, 2020), pp. 343–357. https://doi.org/10.1007/978-3-030-38557-6_16
19. H. HaddadPajouh, A. Dehghantanha, R. Khayami, K.-K.R. Choo, A deep recurrent neural network based approach for Internet of Things malware threat hunting. Futur. Gener. Comput. Syst. **85**, 88–96 (2018). https://doi.org/10.1016/j.future.2018.03.007
20. H. HaddadPajouh, R. Khayami, A. Dehghantanha, K.-K.R. Choo, R.M. Parizi, AI4SAFE-IoT: An AI-powered secure architecture for edge layer of Internet of things. Neural Comput. & Applic. **32**(20), 16119–16133 (2020). https://doi.org/10.1007/s00521-020-04772-3
21. J. Brownlee, A gentle introduction to imbalanced classification (Machine Learning Mastery, 2019). Available online: https://machinelearningmastery.com/what-is-imbalanced-classification/. Accessed 21 July 2020
22. H. Haddadpajouh, A. Azmoodeh, A. Dehghantanha, R.M. Parizi, MVFCC: A multi-view fuzzy consensus clustering model for malware threat attribution. IEEE Access **8**, 139188–139198 (2020)
23. H. Darabian et al., A multiview learning method for malware threat hunting: Windows, IoT and android as case studies. World Wide Web **23**(2), 1241–1260 (2020)

24. A. Yazdinejad, R.M. Parizi, A. Bohlooli, A. Dehghantanha, K.-K.R. Choo, A high-performance framework for a network programmable packet processor using P4 and FPGA. J. Netw. Comput. Appl. **156**, 102564 (2020)
25. Q. Chen, G. Srivastava, R.M. Parizi, M. Aloqaily, I. Al Ridhawi, An incentive-aware blockchain-based solution for internet of fake media things. Inf. Process. Manag., 102370 (2020). https://doi.org/10.1016/j.ipm.2020.102370
26. V. Mothukuri, R.M. Parizi, S. Pouriyeh, Y. Huang, A. Dehghantanha, G. Srivastava, A survey on security and privacy of federated learning. Futur. Gener. Comput. Syst. (2020). https://doi.org/10.1016/j.future.2020.10.007
27. A. Yazdinejad, R.M. Parizi, G. Srivastava, A. Dehghantanha, K.-K.R. Choo, Energy efficient decentralized authentication in internet of underwater things using blockchain, in *2019 IEEE Globecom Workshops (GC Wkshps)*, (2019), pp. 1–6
28. A. Yazdinejad, H. HaddadPajouh, A. Dehghantanha, R.M. Parizi, G. Srivastava, M.-Y. Chen, Cryptocurrency malware hunting: A deep Recurrent Neural Network approach. Appl. Soft Comput. Elsevier **96**, 106630 (2020)
29. M. Aledhari, R. Razzak, R.M. Parizi, F. Saeed, Federated learning: A survey on enabling technologies, protocols, and applications. IEEE Access **8**, 140699–140725 (2020). https://doi.org/10.1109/ACCESS.2020.3013541
30. A. Yazdinejad, R.M. Parizi, A. Dehghantanha, Q. Zhang, K.-K.R. Choo, An energy-efficient SDN controller architecture for IoT networks with blockchain-based security. IEEE Trans. Serv. Comput. **13**(4), 625–638 (2020)
31. A. Yazdinejad, R.M. Parizi, A. Dehghantanha, G. Srivastava, S. Mohan, A.M. Rababah, Cost optimization of secure routing with untrusted devices in software defined networking. J. Parallel Distrib. Comput. (2020). https://doi.org/10.1016/j.jpdc.2020.03.021
32. M. Ask, P. Bondarenko, J.E. Rekdal, A. Nordbø, P. Bloemerus, D. Piatkivskyi, Advanced persistent threat (APT) beyond the hype. Project Report in IMT4582 Network Security at GjøviN University College, vol. 2013 (2013)
33. K. Chang, D.Y.-D. Lin, Advanced persistent threat, p. 12
34. I. Ghafir et al., Detection of advanced persistent threat using machine-learning correlation analysis. Futur. Gener. Comput. Syst. **89**, 349–359 (2018)
35. Y. Su, M. Li, C. Tang, R. Shen, A framework of apt detection based on dynamic analysis, in *Proceedings of the 2015 4th National Conference on Electrical, Electronics and Computer Engineering*, (Xi'an, China, 2015), pp. 1047–1053
36. B. Binde, R. McRee, T.J. O'Connor, Assessing outbound traffic to uncover advanced persistent threat. SANS Institute. Whitepaper, vol. 16 (2011)
37. D. Moon, H. Im, J.D. Lee, J.H. Park, MLDS: Multi-layer defense system for preventing advanced persistent threats. Symmetry (Basel). **6**(4), 997–1010 (2014)
38. G. Vert, B. Gonen, J. Brown, A theoretical model for detection of advanced persistent threat in networks and systems using a finite angular state velocity machine (FAST-VM). Int. J. Comput. Sci. Appl. **3**(2), 63 (2014)
39. M. Belaoued, S. Mazouzi, A Chi-square-based decision for real-time malware detection using PE-file features. J. Inf. Process. Syst. **12**(4), 644–660 (2016)
40. Cyber-research, cyber-research/APTMalware (2020)
41. tf.keras.preprocessing.text.Tokenizer | TensorFlow Core v2.3.0. TensorFlow, https://www.tensorflow.org/api_docs/python/tf/keras/preprocessing/text/Tokenizer. Accessed 14 Aug 2020
42. H. Hashemi, A. Azmoodeh, A. Hamzeh, S. Hashemi, Graph embedding as a new approach for unknown malware detection. J. Comput. Virol. Hacking Tech. **13**(3), 153–166 (2017)
43. J. Bai, Q. Shi, S. Mu, A malware and variant detection method using function call graph isomorphism. Secur. Commun. Netw. **2019**, 1043794 (2019)
44. E. Carrera, pefile: Python PE parsing module
45. World Health Organization, et al., PE – LIEF 0.10.0-845f675 documentation, https://lief.quarkslab.com/doc/stable/api/python/pe.html. Accessed 14 Aug 2020. Osteoarthr. Cartil

46. B. Rocca, Handling imbalanced datasets in machine learning. Medium (2019, March 30), https://towardsdatascience.com/handling-imbalanced-datasets-in-machine-learning-7a0e84220f28. Accessed 14 Aug 2020

47. A. Backurs, P. Indyk, K. Onak, B. Schieber, A. Vakilian, T. Wagner, Scalable fair clustering. arXiv Prepr. arXiv1902.03519 (2019)

48. R. Gandhi, Support vector machine – Introduction to machine learning algorithms. Medium (2018, July 5), https://towardsdatascience.com/support-vector-machine-introduction-to-machine-learningalgorithms-934a444fca47. Accessed 14 Aug 2020

49. P. Gupta, Decision trees in machine learning -towards data science. Towards Data Science (2017), https://towardsdatascience.com/decision-trees-inmachine-learning-641b9c4e8052

Security of Industrial Cyberspace: Fair Clustering with Linear Time Approximation

Nidhip Chikhalia and Yash Dhawan

1 Introduction

Machine learning has unquestionably become one of the most potential and powerful technology around the globe [1]. Its evolution began somewhere in 1950's with Turing test, to the very present intelligent systems [2]. Today, machine learning is used to back the software intended for forecasting, estimation, and analysis [3] Weather forecasting, insurance or mortgage estimation and predictive analytics are only a few examples of where these complex machine learning algorithms meet the needs of mundane users [4–7]. When these algorithms are applied in such real-life scenarios, their precision and fairness needs to be ensured more than ever. Let us consider an Applicant Tracking System (ATS) which automatically filters the job applications at an organization. If the fairness of such system is not maintained, the filtering process can become biased towards a particular group of applicants [8–10]. Now the questions that arise are: Is the process fair? And Is the outcome fair? The answers to these questions can only be justified by the algorithm running behind the process. So, [11] introduces a fair variant of the classic or as mentioned in their work, vanilla version of the K-median clustering problem. They introduce a technique that makes sure that a protected or sensitive class has an approximately equal representation in each cluster. The method is divided into two phases, where the first phase divides the input pointset into small subsets called fairlets. The input points are assigned colors and these fairlets maintains the balance of each color. In the second phase, these fairlets are merged to form clusters and for this clustering, an existing algorithm is used such as K-center or K-median clustering. Here, the fairlet decomposition phase is the main part which ensures the fairness component. The

N. Chikhalia (✉) · Y. Dhawan
School of Computer Science, University of Guelph, Guelph, ON, Canada
e-mail: nchikhal@uoguelph.ca; ydhawan@uoguelph.ca

© Springer Nature Switzerland AG 2021
K.-K. R. Choo, A. Dehghantanha (eds.), *Handbook of Big Data Analytics and Forensics*, https://doi.org/10.1007/978-3-030-74753-4_5

only problem faced here is the fairlet decomposition time, which is super-quadratic compared to the number of input points.

To address this issue, [12] introduces a new method for fairlet decomposition. This new method aims to reduce the fairlet decomposition time nearly liner to the number of input points [13–18]. For this, they introduce an HST function which is fed with the input points. This HST function forms a tree structure where each node of the tree is an input point. The fairlet decomposition is then done using the distance obtained from the HST tree and not the actual input points. The algorithm then follows the same procedure as the original one, i.e. clustering the fairlets using existing K-median algorithm. Both of these works demonstrate their performance on the general datasets like census, diabetes, and bank. Our aim in this paper is to test the algorithm developed by [11] and improved by [12] on the datasets used to train the machine learning models for cybersecurity. These datasets are IoT Malware Dataset and Industry Control System datasets like Secure Water Treatment and BATADAL. Our contribution to this paper is a mechanism that compares the output of the clustering algorithm with the original dataset in order to confirm the correctness of the results obtained.

This work includes a literature review of a few related articles, methodology which describes about datasets, algorithm, and our experiment in detail and then the results obtained, and the comparisons done with the original work.

2   Literature Review

Industrial Control System (ICS) refers to a wide range of systems the monitors and measures the control and also automate the processes in a broad range of industries [19, 20]. ICS are of various types, such as, Distributed Control System (DCS), Supervisory Control and Data Acquisition Systems (SCADA), Programmable Logic Controllers (PLCs) and Safety Instrumented Systems (SIS) [21, 22]. Some of these systems are legacy systems and were developed decades ago when the cyber threat was not as big of a threat. But in today's world, cyberattack are a huge threat to ICS, and the ICS being crucial for the industry sector, it is very important to secure it from the cyber threats [23].

Research is being conducted on using Artificial Intelligence and Machine learning to detect and mitigate these cyber risks in the Industrial Control Systems, [24] gives an in depth explanation of how the evolution of industrial control system has exposed it to the new threats of cyberattacks and how these attacks or unintentional mistakes can affect an infrastructure that is serving a huge number of people. The researchers here demonstrates the work on VERIS community database, and the motivation is to use the incidents described in this database to gain a better understanding of the cyber threats on the industry. Using he Monte Carlo simulation, the researchers aim to predict the possibility of these attacks in the future, and also aim to make this process repeatable in order to keep predicting the threats repeatedly for the future.

Karimipour and Leung [25] demonstrates a work on securing a Cyber Physical System (CPS) like power grid from cyberattacks. The authors here specifically targets anomaly detection and the attack vector know has False Data Injection (FDI) where the attacker injects false data into the system such that it cannot be detected by the conventional anomaly detection systems. The work here describes the use of EnKF model to predict the state of the system. EnKF uses historic data to predict the state, the existing predictors use measurement of the system to predict the state which may not be secure enough in case of FDI. The predictor model proposed here uses the history of the system to predict the state and identify the anomaly.

Another work [26] describes the methods of detecting crypto-ransomware in the Internet of Things network. The authors proposes a model that monitors the power consumption of the devices on the network and uses these consumption trends to determine if the application is a ransomware or not. The authors use power tutor to monitor and sample this consumption data from the android applications like Gmail, Facebook, Google Chrome, YouTube, WhatsApp, Skype and six other recent and active ransomware applications. The result of the proposed model is claimed by the authors to be better than KNN, Neural Networks, SVM and Random Forest.

Karimipour et al. [27] is a work that describes an unsupervised machine learning model that detects the cyberattacks in the large scale physical cyber space. The proposed model aims at a scalable anomaly detection mechanism that uses statistical correlation between the measurements in a smart power grid. Symbolic Dynamic Functioning is used for feature extraction in order to reduce the computational cost. The aim of the research is justified by the results obtained which shows 99% accuracy of anomaly detection and 98% true positive rate. This work demonstrates an unsupervised machine learning model which unlike previous works does not use historical or previously obtained data for training.

Focusing on the works related to clustering or the fairness of the clustering, a work by [28] focuses on improving the algorithm introduced by [11] They introduce a concept of coresets (S), such that S is a subset of input pointset P ($S \subset P$). Basically, the aim of the algorithm is to compute the coreset in a near linear time and running the k -median clustering on this coreset would yield the approximate solution for the original input points. The approach implementing HST function by [12] can be considered complementary to the coreset computation. Also, [12] states that both concepts can be applied together in order to reduce the computation time and also the required memory.

Another contribution by [29] works on enforcing the fairness component to the correlation graph clustering algorithm. The high-level aim of the paper is to minimize the disagreements among the vertices of the graph in the same cluster. For this they propose two techniques, one is similar to the techniques proposed by [11], that is assigning colors to the vertices and ensuring that any cluster is not dominated by a single color. The second technique is to set relative upper and lower bounds for the number of vertices of any color in a cluster. Since it is a graph and not a metric space, there is only one phase. There is no decomposition of the graph into subgraphs. Initially it demonstrates the results with only two colors, but later they prove with their results that the algorithm runs perfectly for more than two colors, that is more than two values of a compulsory attribute.

Bera et al. [30] again generalizes the work of [11] Here generalization means that the algorithm works on l_p-norm objective (k-means, k-median and k-center), the only drawback is a little loss in quality of output compared to the original work. The proposed work allows the user to specify the sensitive attribute and also the upper and lower bounds of the sensitive attribute in each cluster. The authors also claim that the algorithm can perform without the sensitive attribute and still maintain the fairness. But the fairness notion here is a little different than the original work, here the algorithm allows more than one protected group and also the overlapping of those groups. Also, the algorithm works opposite to the original work, in first phase the vanilla clustering is applied, and the clusters are formed. Then in the second phase, the points in the clusters are analyzed and re-positioned to obtain fairness. The algorithm is tested on the same dataset as the original work and even though the work adds a lot of functionality to the existing algorithm, the results obtained are similar but not better.

3   Methodology

(a) **Dataset**

This section gives a brief on the datasets used by us for our experiment. We used three algorithms namely IoT malware dataset, Secure Water Treatment dataset and BATADAL.

IoT Malware Dataset
The IoT Malware dataset has been constructed using the machine opcodes. It is comparatively small dataset and contains only 513 instances. Out of those 513, 269 indicate the machine behavior when it is operating under normal conditions and rest 245 indicate the machine under attack state. Each instance has a feature set of 236 features.

SWaT
Secure water treatment (SWaT) dataset is constructed using the data collected from 11 days of continuous operation of water treatment testbed. Out of these 11 days, for 7 days, the system was operating under normal conditions and for the rest of 4 days the data collected was for the system conditions under attack. Historian was used to store all the data collected from all network traffic, sensors, and actuator data. The dataset consists of about 15,000 datapoints where each datapoint has a feature set of 78 features. Of these 15,000 datapoints, around 9500 depict the environment under normal conditions and rest describes the environment under attack conditions [16].

BATADAL
BATADAL was constructed for Battle of Attack Detection Algorithms which is a competition to compare the performance of the cyber-attack detection algorithms

in water distribution systems. This dataset is not based on real life data but is considered as realistic because it was constructed using the data collected from the de facto standard water distribution simulation tool namely EPANET. The dataset consists of roughly 13,000 datapoints where 12,500 are normal scenarios and rest 500 are attack scenarios. Each datapoint has a feature set of 46 attributes [31].

(b) **Feature Selection**

Since all these datasets have a huge feature set and very little knowledge about these features is available, it is hard to depict what is the function or purpose of a feature. For our algorithm, we need a sensitive feature. [The base paper] uses gender as the sensitive attribute from the diabetes dataset. Here we perform feature selection on each dataset using Extra Tree Classifier to derive the sensitive attribute as well as the best features that we can use to improve the performance of the algorithm.

Figure 1 on the left shows the top features for IoT Malware dataset. From this dataset, for our experiment, we have used total of 7 features. Here we have used 'mov' as the sensitive feature and 'label' as the final or the seventh attribute. The significance of the final attribute here is that it is used to measure the performance of the algorithm.

Figure 2 on the left depicts the top 10 features for SWaT dataset. Here we have considered top 6 features, where 'AIT 201' is considered as the sensitive attribute and 'label' is used as the final attribute for comparison with the predicted labels (Fig. 3).

(c) **Brief description of the Algorithm**

The fair variant of the classic K-means algorithms introduced by [11] takes an input pointset P, the points $p \subset P$ are assigned colors based on the sensitive attribute

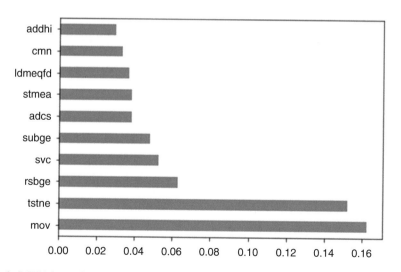

Fig. 1 IoT Malware dataset

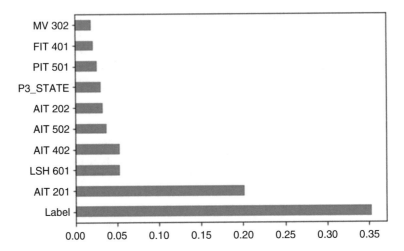

Fig. 2 SWaT dataset

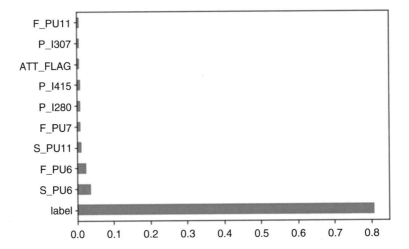

Fig. 3 BATADAL dataset

in the pointset (For example Red and Blue). The algorithm in its first phase divides the input points into small clusters preserving the balance of each colored input point in the cluster. The method used for the division of these input points affects the approximation guarantees of the final clustering algorithm.

The balance being talked about here can be defined as:

For input pointset p, balance will be

$$\text{Balance } (p) = \min \ (\#\text{Red} (p) / \#\text{Blue} (p) , \#\text{Blue} (p) / \#\text{Red} (p)) \in [0, 1]$$

A subset p of P is said to be perfectly balanced if it has equal number of reds and blues, balance (p) =1, while it is said to be completely unbalanced if balance (p) = 0, where the subset is monochromatic. This phase is named as the fairlet decomposition and the small clusters are mentioned as the fairlets.

In the second phase, these fairlets are clustered using the function (t, k) – fair center or (t, k) – fair median where 't' is the balance discussed above and 'k' is the number of clusters C.

The time taken by the first phase of this algorithm is quadratic compared to the number of inputs [12]. introduces a method of fairlet decomposition to make the running time near-linear to the number of input points. This again divides the first phase of the algorithm, i.e. fairlet decomposition into two parts, first it inputs the pointset into a tree metric called HST. Here a quadtree computation of the input points is done. In the second part the fairlet decomposition is done using the distances in the quad tree and not the actual input points. Then it follows the same procedure as the second phase of the algorithm to cluster the fairlets. Here it only uses the (t, k) – fairlet median clustering.

The most important step here is embedding the input pointset into the HST tree and constructing the γ-HST tree. For the γ-HST tree to be well separated, it is expected to follow two conditions:

(i) The weighted distance from any node in the tree to its children are same.
(ii) For each node v, the distance between v and its children should be at most $1/\gamma$ times the distance between v and its parent.

Once the HST tree is constructed, it is parsed in a top down manner to create fairlets. To partition the points into (r, b) – fairlets from the nodes it is assumed that each node v is augmented with extra information like Number of Red (N_r) and Number of Blue (N_b) points. The Fig. 4 shows the high-level working of the algorithm for (1,3) – fairlet decomposition with 8 blue and 4 red points.

Our contribution to this work is that we implement an existing method to verify the precision of the algorithm by comparing the result of the clustering with the original data. For this, we embed an attribute called 'label' that indicates if the input point describes a normal or an attack environment in the system to our dataset. It is also of importance that the algorithm doesn't accept string inputs, hence this indicating attribute is mapped to integers. General mapping is that the string 'Normal' is mapped to '0' and the string 'Attack' is mapped to '1'. The position of this attribute in the dataset is obtained manually and is hard coded in order to create a reference for the comparison. In order to automate the process of acquiring this attribute, we need to make sure that this attribute forms the last column of the dataset.

Since there are only 2 groups, i.e. Attack and Normal, we assign each of the clusters to one of the groups. For simplicity, let's assume k = 2. This means that the dataset will be clustered into 2 clusters, we expect one cluster to be Normal and the other one to be Attack and assign each cluster to a different group. Using the actual and the predicted labels, we create a confusion matrix which gives out the precision, recall and f1 score of the algorithm for that dataset.

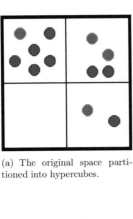

(a) The original space partitioned into hypercubes.

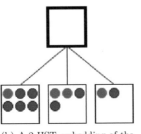

(b) A 2-HST embedding of the input points.

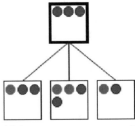

(c) Stage 1: we must connect 3 blue points from the left node through the root.

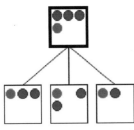

(d) Stage 2: we can connect 1 red point from the middle node through the root.

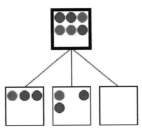

(e) Stage 3: we add the unsaturated fairlet in the right node to the root and make it balanced.

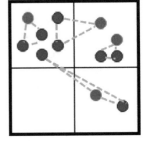

(f) The final fairlet clustering.

Fig. 4 Overview of the algorithm

(a) Experiments

Our experiment include the reimplementation of algorithm produced by [12] on the diabetic dataset mentioned in the work. We also implemented the algorithm on three other datasets, IoT malware dataset, Secure Water Treatment dataset (SWaT) and Battle of Attack Detection Algorithm (BATADAL).

The original work by [12] implements the algorithm using 2 dimensions from the diabetic dataset, the dimensions being 'age' and 'time in hospital'. The sensitive attribute here is 'gender'.

In our implementation of the paper, we used the same dimensions and sensitive attribute and divided into 20 clusters to obtain the similar results. Then we implemented the algorithm with the same dataset and sensitive attribute but different dimensions.

After experimenting with the diabetic dataset, we ran the algorithm on the IoT Malware dataset. For this we used 7 attributes and one sensitive attribute and divided it into 6 clusters. The 7 attributes used by us here are the top 6 features obtained during the feature selection and the 7th attribute is 'label' which is used as a reference to measure the precision of the algorithm. The sensitive attribute here is the topmost feature 'mov'.

After IoT Malware dataset, we conducted the experiment on SWaT dataset. Here we considered top 5 features from the feature selection and 'label' as the 6th attribute. We divided the algorithm into 2 clusters, assuming one would be 'normal' and the other would be 'attack'. The sensitive attribute here was 'AIT 201'.

The final dataset we used for testing the algorithm was BATADAL, which we clustered into 4 clusters. The number of dimensions here were 7 and one sensitive attribute being 's_PU6'. The setting for the attributes was the usual, first 6 attributes being the top 6 from the feature selection and the last one being 'label'.

The results obtained from these experiments are discussed in the next section.

4 Results and Conclusion

The goal of the algorithm introduced by [12] was to reduce the fairlet decomposition time to near linear with respect to the number of input points. Our experiment to reimplement the same algorithms on the same dataset with the same settings gave us the results shown below in Figs. 5 and 6. As seen in the Fig. 6, the time for fairlet decomposition increases linearly with the increase in number of data points which initially was increasing quadratic to the number of input points.

This result confirms that the environment setup by us for the reimplementation of the algorithm was functioning correctly and the desired results were obtained. The p:q ratio for this experiment was 4:5 and the number of clusters made here

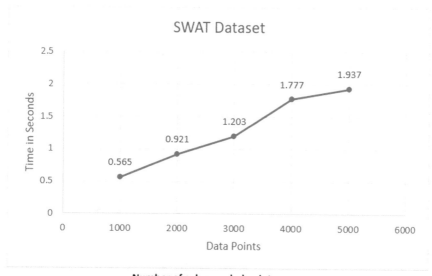

Fig. 5 Original results

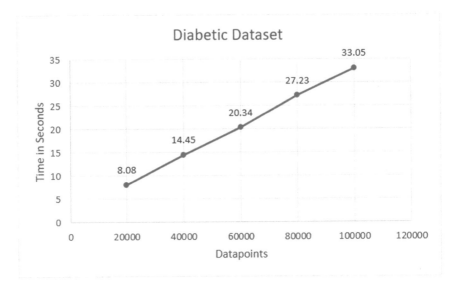

Fig. 6 Results of reimplementation

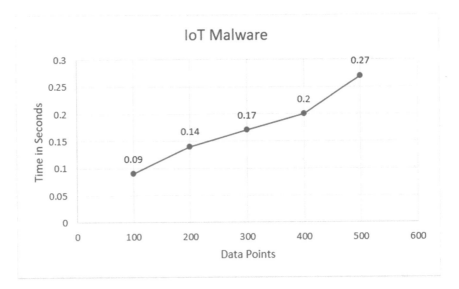

Fig. 7 IoT Malware dataset time

were 20. After this confirmation, we could proceed experimenting with the other datasets. The next experiment was on IoT malware dataset. The results we obtained are shown below in Figs. 7 and 8. The p and q values here were chosen as 1 and 5 respectively and the dataset was divided into 6 clusters.

	precision	recall	f1-score	support
0	0.54	0.85	0.66	40
1	0.77	0.41	0.53	49
avg / total	0.67	0.61	0.59	89

Fig. 8 IoT Malware dataset performance

	precision	recall	f1-score	support
0	0.90	1.00	0.95	1190
1	1.00	0.76	0.86	559
avg / total	0.93	0.92	0.92	1749

Fig. 9 SWaT dataset performance

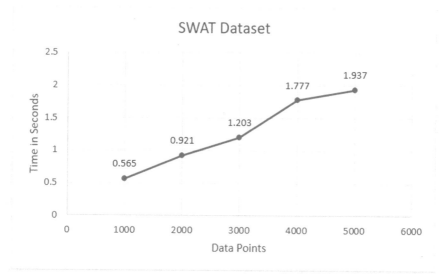

Fig. 10 SWaT dataset time

The results show that the time for fairlet decomposition of IoT Malware dataset is not exactly linear with the number of datapoints by is near to linear. The Fig. 8 shows the performance of the algorithm over this dataset, which is not excellent but fairly good. Next, we experiment with the SWaT dataset. Dividing it into just two clusters, assuming one would be 'Normal' and the other would be 'Attack' environment. The p and q balance obtained for this dataset was 1 and 8 respectively. The results obtained were excellent and are indicated in the Fig. 9. Figure 10 depicts the Time vs Number of Datapoints graph which is almost linear.

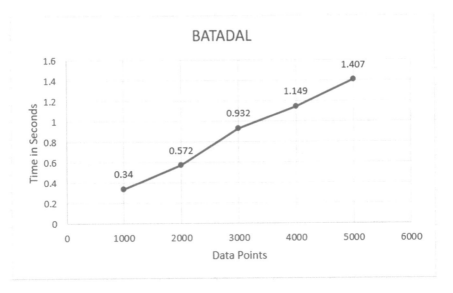

Fig. 11 BATADAL dataset time

	precision	recall	f1-score	support
0	0.95	0.84	0.89	99
1	0.00	0.00	0.00	4
avg / total	0.92	0.81	0.86	103

Fig. 12 BATADAL dataset performance

The final experiment we carried out was on the BATADAL dataset. This dataset being artificially synthesized, was expected to give very good results. It was clustered into 4 different clusters, where we assigned 2 clusters to 'Normal' and remaining two clusters to 'Attack' scenario. The difficult part here was to obtain the p and q balance which was achieved at 1 and 125 respectively. The Fig. 11 shows the most liner time approximation obtained among all the new datasets and the Fig. 12 shows the overall performance of the algorithm over this dataset which is one of the best one.

Hence it can be said that algorithm shows the best performance over the BATADAL dataset. The Table 1 below gives the summary of the results obtained from all the datasets and the comparison with the results from the original work. It also mentions the Fairlet Decomposition Cost and the K-median Clustering cost of each dataset.

From here, it can be concluded that the work done by [12] with an aim to reduce the fairlet decomposition time to almost linear works for multiple algorithms. The K-median clustering followed by it, is accurate and yields excellent results for the datasets used to train and test the machine learning model for cybersecurity

Table 1 Final results

Dataset	Datapoints	p, q	k	K-median cost	Fairlet decomposition cost
Diabetes (by [12])	1000	4, 5	20	4149	2971
Diabetes (by us)	1000	4, 5	20	28,531	12,248
IoT Malware	500	1, 5	6	50.11	49.72
SWaT	5000	1, 8	2	49272.88	15650.23
BATADAL	5000	1, 125	4	186899.29	18407.95

purposes. Hence such algorithm can be used for securing the Industry Control Systems' cyberspace and by developing more accurate datasets like BATADAL, the systems can be trained even better which provides another level of security.

References

1. S.M. Tahsien, H. Karimipour, P. Spachos, Machine learning based solutions for security of Internet of Things (IoT): A survey. J. Netw. Comput. Appl. **161**, 102630 (2020)
2. V. Sharma, The exciting evolution of machine learning. Vinod Sharma's Blog (2018), https://vinodsblog.com/2018/03/11/the-exciting-evolution-of-machine-learning/. Accessed 23 Apr 2020
3. H. HaddadPajouh, A. Dehghantanha, R.M. Parizi, M. Aledhari, H. Karimipour, A survey on internet of things security: Requirements, challenges, and solutions. Internet of Things, 100129 (2019). https://doi.org/10.1016/j.iot.2019.100129
4. A. Yazdinejad, G. Srivastava, R.M. Parizi, A. Dehghantanha, H. Karimipour, S.R. Karizno, SLPoW: Secure and low latency proof of work protocol for blockchain in green IoT networks, in *2020 IEEE 91st Vehicular Technology Conference (VTC2020-Spring)*, vol. 2020, pp. 1–5
5. A. Yazdinejad, R.M. Parizi, A. Dehghantanha, K.-K.R. Choo, Blockchain-enabled authentication handover with efficient privacy protection in SDN-based 5G networks. IEEE Trans. Netw. Sci. Eng. (2019). https://doi.org/10.1109/TNSE.2019.2937481
6. A. Singh, K. Click, R.M. Parizi, Q. Zhang, A. Dehghantanha, K.-K.R. Choo, Sidechain technologies in blockchain networks: An examination and state-of-the-art review. J. Netw. Comput. Appl. **149**, 102471 (2020). https://doi.org/10.1016/j.jnca.2019.102471
7. A. Yazdinejad, R.M. Parizi, A. Dehghantanha, Q. Zhang, K.-K.R. Choo, An energy-efficient SDN controller architecture for IoT networks with blockchain-based security. IEEE Trans. Serv. Comput. **13**(4), 625–638 (2020)
8. D. Połap, G. Srivastava, A. Jolfaei, R.M. Parizi, Blockchain technology and neural networks for the internet of medical things, in *IEEE INFOCOM 2020 – IEEE Conference on Computer Communications Workshops (INFOCOM WKSHPS)*, (2020), pp. 508–513. https://doi.org/10.1109/INFOCOMWKSHPS50562.2020.9162735
9. A. Yazdinejad, G. Srivastava, R.M. Parizi, A. Dehghantanha, K.-K.R. Choo, M. Aledhari, Decentralized authentication of distributed patients in hospital networks using blockchain. IEEE J. Biomed. Heal. Inform. **24**(8), 2146–2156 (2020)
10. Q. Chen, G. Srivastava, R.M. Parizi, M. Aloqaily, I. Al Ridhawi, An incentive-aware blockchain-based solution for internet of fake media things. Inf. Process. Manag., 102370 (2020). https://doi.org/10.1016/j.ipm.2020.102370
11. F. Chierichetti, R. Kumar, S. Lattanzi, S. Vassilvitskii, Fair clustering through fairlets, in *Advances in Neural Information Processing Systems*, (MIT Press, Cambridge, MA, 2017), pp. 5029–5037

12. A. Backurs, P. Indyk, K. Onak, B. Schieber, A. Vakilian, T. Wagner, Scalable fair clustering. arXiv Prepr. arXiv1902.03519 (2019)
13. A. Yazdinejad, R.M. Parizi, A. Bohlooli, A. Dehghantanha, K.-K.R. Choo, A high-performance framework for a network programmable packet processor using P4 and FPGA. J. Netw. Comput. Appl. **156**, 102564 (2020)
14. R.M. Parizi, S. Homayoun, A. Yazdinejad, A. Dehghantanha, K.-K.R. Choo, Integrating privacy enhancing techniques into blockchains using sidechains, in *IEEE Canadian Conference of Electrical and Computer Engineering, CCECE 2019*, (2019). https://doi.org/10.1109/CCECE.2019.8861821
15. A. Yazdinejad, R.M. Parizi, G. Srivastava, A. Dehghantanha, K.-K.R. Choo, Energy efficient decentralized authentication in internet of underwater things using blockchain, in *2019 IEEE Globecom Workshops (GC Wkshps)*, (2019), pp. 1–6
16. V. Mothukuri, R.M. Parizi, S. Pouriyeh, Y. Huang, A. Dehghantanha, G. Srivastava, A survey on security and privacy of federated learning. Futur. Gener. Comput. Syst. **115**, 619–640 (2020)
17. A. Yazdinejad, H. HaddadPajouh, A. Dehghantanha, R.M. Parizi, G. Srivastava, M.-Y. Chen, Cryptocurrency malware hunting: A deep recurrent neural network approach. Appl. Soft Comput. J. Elsevier **96**, 106630 (2020)
18. M. Aledhari, R. Razzak, R.M. Parizi, F. Saeed, Federated learning: A survey on enabling technologies, protocols, and applications. IEEE Access **8**, 140699–140725 (2020). https://doi.org/10.1109/ACCESS.2020.3013541
19. A. Al-Abassi, H. Karimipour, A. Dehghantanha, R.M. Parizi, An ensemble deep learning-based cyber-attack detection in industrial control system. IEEE Access **8**, 83965–83973 (2020)
20. A. Yazdinejad, R.M. Parizi, A. Dehghantanha, K.-K.R. Choo, P4-to-blockchain: A secure blockchain-enabled packet parser for software defined networking. Comput. Secur. **88** (2020). https://doi.org/10.1016/j.cose.2019.101629
21. H. Karimipour, V. Dinavahi, Extended Kalman filter-based parallel dynamic state estimation. IEEE Trans. Smart Grid **6**(3), 1539–1549 (2015)
22. A. Yazdinejad, A. Bohlooli, K. Jamshidi, Efficient design and hardware implementation of the OpenFlow v1.3 Switch on the Virtex-6 FPGA ML605. J. Supercomput. **74**(3), 1299–1320 (2018). https://doi.org/10.1007/s11227-017-2175-7
23. H. Karimipour, V. Dinavahi, Robust massively parallel dynamic state estimation of power systems against cyber-attack. IEEE Access **6**, 2984–2995 (2017)
24. S. Walker-Roberts, M. Hammoudeh, O. Aldabbas, M. Aydin, A. Dehghantanha, Threats on the horizon: Understanding security threats in the era of cyber-physical systems. J. Supercomput. **76**(4), 2643–2664 (2020)
25. H. Karimipour, H. Leung, Relaxation-based anomaly detection in cyber-physical systems using ensemble Kalman filter. IET Cyber-Phys. Syst. Theory Appl. **5**(1), 49–58 (2020)
26. H. Darabian et al., Detecting cryptomining malware: A deep learning approach for static and dynamic analysis. J. Grid Comput. **18**, 1–11 (2020)
27. H. Karimipour, A. Dehghantanha, R.M. Parizi, K.-K.R. Choo, H. Leung, A deep and scalable unsupervised machine learning system for cyber-attack detection in large-scale smart grids. IEEE Access **7**, 80778–80788 (2019)
28. M. Schmidt, C. Schwiegelshohn, C. Sohler, Fair coresets and streaming algorithms for fair k-means clustering. arXiv Prepr. arXiv1812.10854 (2018)
29. S. Ahmadi, S. Galhotra, B. Saha, R. Schwartz, Fair correlation clustering. arXiv:200203508 [cs, stat] (2020)
30. S. Bera, D. Chakrabarty, N. Flores, M. Negahbani, Fair algorithms for clustering, in *Advances in Neural Information Processing Systems*, (MIT Press, Cambridge, MA, 2019), pp. 4954–4965
31. Q. Lin, S. Verwer, R. Kooij, A. Mathur, Using datasets from industrial control systems for cyber security research and education, in *International Conference on Critical Information Infrastructures Security*, (2019), pp. 122–133

Adaptive Neural Trees for Attack Detection in Cyber Physical Systems

Alex Chenxingyu Chen and Kenneth Wulff

1 Introduction

The race to get machines or computers to think and behave like humans have gained traction over the years, and this has provoked tremendous advancements in the field of Artificial Intelligence (AI) [1–3] and Machine learning (ML) [4–6]. The human brain is made up of neurons that are all connected to form a very complicated internetwork of brain cells, and neural networks are designed to mimic the way the brain works. Decision trees primarily build a tree structure to model a set of sequential or hierarchical decisions that eventually lead to an outcome.

Even though artificial intelligence and machine learning are usually used interchangeably, there is a difference between them. Artificial intelligence focuses on the replication of human intelligence in computers whiles machine learning deals with the ability of a machine to learn using large datasets instead of relying on hard-coded rules. Machine learning enhances the ability of computers to self-learn. The two work together to improve the intelligence of the machine and to make them behave and act more like humans [7].

The McKinsey Global Institute (MCI) reports that artificial intelligence investments are growing fast. Furthermore, it estimates that technology giants globally spent between USD 20 billion to USD 30 billion on artificial intelligence investments in 2016. The MCI notes that machine learning as an enabling technology received the largest share of this investment. The institute purports that machine learning adoption outside technology is in its infancy and that this presents a massive opportunity for growth in the machine learning field going forward [8–10].

A. C. Chen (✉) · K. Wulff
School of Computer Science, University of Guelph, Guelph, ON, Canada
e-mail: cchen22@uoguelph.ca; kwulff@uoguelph.ca

© Springer Nature Switzerland AG 2021
K.-K. R. Choo, A. Dehghantanha (eds.), *Handbook of Big Data Analytics and Forensics*, https://doi.org/10.1007/978-3-030-74753-4_6

Marketandmarket reveals that the machine learning space is set to grow from USD 1.03 billion to 8.81 billion from 2016 through 2022, with a compound annual growth rate of 44.1%. They state that the application of machine learning in various industry verticals is set to increase exponentially as more and more companies and industries adopt machine learning in their day to day operations. Marketandmarket infers that advancements in technology and the proliferation of data generated in various market segments account for some of the factors driving growth in machine learning usage [11].

BCCResearch also states that the global machine learning market totalled USD 1.4 billion in 2017, and it is projected to reach USD 8.8 billion by 2022 at a compound annual growth rate of 43.6%. The report concludes based on analysis of the machine learning vendor landscape and the profiles of the major players in the global machine learning market that adoption and use of machine learning are going to increase year on year because of its adoption in many different solutions and services spanning all industry vertical from energy, healthcare, finance and telecommunications to the military [12–15].

Businesswire reports that according to the latest research, the machine learning market is set to grow by USD 11.16 from 2020 to 2024. The report explains that the rising adoption of cloud computing services globally in different multi-user industries account for this growth. Businesswire states that the many benefits of cloud computing, such as the minimal cost of computing operations, scalability, reliability, and high resource available, encourage many enterprises to transition to cloud computing and indirectly adopt machine learning via these machine learning-enabled cloud services. Therefore, these offerings have become primary factors that are key to driving the global machine learning market in the years to come [16].

The rise in machine learning adoption and patronage is primarily due to the many benefits companies continue to accrue from the use and deployment of machine learning in their operations. This growth has spurred the need for advancement and to look for new machine learning methods that are novel and improve on the traditional methods to deliver outstanding results as needs get more complex [5, 17].

In this research, we evaluate four different datasets using the adaptive neural trees approach to determine its accuracy against these datasets. Our objective is to ascertain, which of the datasets would generate the best outcome when processed through the adaptive neural trees algorithm.

To understand this research, the reader needs a good understanding of machine learning and the traditional methods available in use today because our approach pulls from the strengths of some of these conventional methods to deliver a much more superior outcome.

The next sections of this paper discuss the literature review, methodology, results and discussion and conclusion and future work. The literature review addresses other people's work in this domain. The methodology outlines the various steps and processes we undertook during the research to arrive at our answers. The results of our research are detailed and reviewed in the results and discussion portion. We discuss the conclusion and future work in the final section of the paper.

2 Literature Review

Advancing and improving the frontiers of machine learning in this big data era have become imperative as more and more businesses incorporate machine learning techniques and practices into their everyday operations [18–24]. This need has given rise to research that is geared towards improving the existing methodologies to meet the increasing need for more accurate results from complex datasets. To this end, our research was conducted to showcase the benefits of aggregating the strengths of Decision Trees (DT) and Neural Networks (NN) to optimize algorithmic results from the given datasets [25–28] We do this by looking at previous work in the field and highlight the limitations in these previous studies.

Decision trees, which are usually referred to as Classification and Regression trees, is one of the predictive modelling methods used in machine learning. Nevertheless, they are not the best algorithms for image classification because they can cause overfitting. It uses a tree-like model to predict and go from observations about an item to conclusions about the value of the target. Even with pruning, they still do not deliver the best results when dealing with vast datasets of images. Decision trees also suffer from variance because a small variation in the dataset can result in a completely different tree being generated. Also, they can become non-robust such that a small change in the training data could result in a significant change in the tree structure and consequently affect the final prediction and render the results unreliable for use in machine learning services [29].

Gradient Boosting Decision Trees (GBDTs), however, have been extensively used for image labelling and in advertising systems. Training GBDTs is ineffective and very time consuming, especially when there are deep trees and large datasets involved. In 2018, Zei Wen et al. proposed a novel Graphics Processing Unit (GPU) based algorithm called GPU-GBDT to enhance GBDT training. Whiles this was great work, it had one major limitation. Their research improved efficiency compared to existing techniques. However, it did not address the issue of accuracy, which is critical in improving processes that rely on advancements in machine learning to deliver essential services across many industries [30].

Feature extraction to transform unstructured data to structured data sometimes presents large dimensionality, which could contain large amounts of irrelevant features increasing the computational complexity for the learning algorithm, which sometimes leads to overfitting of the training dataset [31–33]. Because most algorithms are susceptible to irrelevant features, it has become crucial to efficiently evaluate features and choose only the relevant ones for the learning models if accuracy is of the essence [34].

Han Liu et al. in 2017, showed that this could be achieved by using two approaches called filter and wrapper. The filter approach evaluates features before the training stage and selects a subset of the features for the learning model. One drawback of the filter method is that it assumes that features with a higher variance may contain more useful information, which is usually not the case. The filter method also does not consider the relationship between feature variables or feature and target variables, which is another drawback of the filter approach. The wrapper methodology uses an algorithm to discover models from various feature subsets and compare the analytical performance of the models to evaluate the resulting subset features. The limitations of this approach are that they are discriminative and require very high computational intensity. Also, they are prone to a high risk of overfitting, which eventually affects their efficiency and accuracy when an enormous volume of datasets is thrown at it [35].

Deep Neural Networks (DNNs) have transformed the machine learning field. They are now in use in many applications across many industries, such as computer vision, where it has delivered some fantastic results on a plethora of challenges. However, these successes have been on the back of training data images that have been familiar to classify. These models, however, falter in their ability to correctly classify data when they are presented with unfamiliar objects. Unfamiliarity could stem from images that have a different orientation, brightness, colour, or scale, just to name a few differentiating attributes.

In December 2017, Hossein et al. were able to demonstrate that when the "semantic generalizations" of the images were altered, the models were unable to decipher and categorize images accurately. Their experiment was conducted on standard image datasets such as MNIST and the modified VGG networks trained on colour and grayscale variations of the German Traffic Sign Recognition Benchmark (GTSRM) and the CIFAR-o dataset. They showed that altering the familiarity or semantic generalizations of the datasets produced accurate rates that were significantly lower than regular images, and that the results were relatively good only when there was a significant range of variety within the training set.

Their position was that neural networks sometimes underperform because test data is not distributed in the same way as training data, which is what happens in the real world. Because of this, the models are unable to learn the structure of the objects in question effectively and are, therefore, incapable of semantically differentiating between the different object classes to compute an accurate prediction [36].

In our research, we were able to eliminate this problem by using the ImageFolder function to read and resize the images, and by using transformers on the IoT datasets. For the BATADAL and SWAT datasets, we avoided this conundrum by "Upsampling" the dataset and by using the Imbalanced Dataset Sampler to balance the dataset before training the models.

Our research improves both the efficiency and the accuracy of existing models to deliver improved results whiles eliminating the difficulties associated with most of the previous research papers [37–44]. As earlier stated, it also handles noise and inaccuracies connected to other models by Upsampling and balancing the datasets before training the models. It is simple and straightforward to set up and replicate the results without the need for any dedicated or specialized hardware. One can spin up instances in the Google Cloud Platform (GCP) or any other cloud platform and conduct this experiment at a little cost to the researcher but with phenomenal results.

3 Methodology

In the methodology section, we present the data processing and training workflow for reproducing the author's result as well as results from the IoT [45, 46], BATADAL and SWAT datasets. Figure 1 shows an overview of the IoT dataset methodology. Figure 2 shows an overview of both BATADAL and SWAT dataset methodologies.

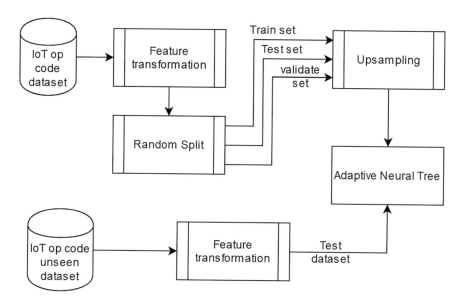

Fig. 1 IoT dataset methodology

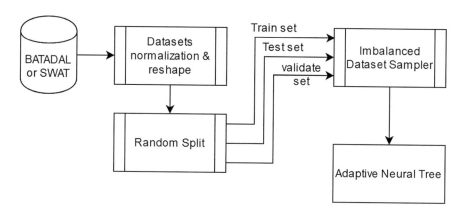

Fig. 2 BATADAL and SWAT dataset methodology

3.1 Environment Setup and Dataset Download

The system used to perform the analysis was a GPU virtual machine in the Google Cloud Platform. The operating system was Ubuntu 16.04 with four cores of virtual CPU and an NVIDIA Tesla P100 GPU and 15 GB of RAM.

We used the Anaconda 2020.02 Linux version to set up the Python 2.7 environment with CUDA 8.0 and PyTorch 0.3.0. Anaconda is an all-in-one installer, which includes all the necessary python packages for data science. We used the IoT malware, BATADAL and SWAT datasets from the Cyber Science lab.

3.2 Reproducing Author's Results

R. Tanno et al. proposed a new machine learning algorithm called Adaptive Neural Trees (ANT) [47], which we used to detect IoT malware and for classifying the BATADAL and SWAT attacks.

Because the author had already provided the configuration details and how to train the built-in MNIST dataset, we proceeded to train the model with the author's configuration and observed the results. Table 1 shows the configuration the author used to produce the results in the base paper.

3.3 Adding Functions to ANT

In order to generate the accuracy, true/false positive rates, true/false negative rates, F1 score, precision, and recall, we changed the log code to make sure it could record the confusion matrix when it validates the model.

Table 1 Author's configuration

Dataset	MNIST
Batch-size	256
Epoch patience	5
Epoch node	50
Epoch finetune	100
Transformer	$1 \times$ Conv3-5* $+ 1 \times$ MaxPool
Router	$1 \times$ Conv3-5 $+$ GAP $+ 2 \times$ FC
Solver	GAP $+ 2$ FC layers $+$ Softmax
Downsample frequency	1
Max depth	10
Randomization seed	1

*Conv3-5 means a 2D convolution with 3 kernels of spatial size 5×5
GAP stands for Global-Average-Pooling, FC stands for Fully Connected layer

3.4 Dataset Preprocessing

This section includes all the data processing details for the three datasets. Because BATADAL and SWAT datasets share the same dataset structure, they will be put into one subsection.

3.4.1 IoT Dataset

The IoT dataset file contained benign and malicious operation code text files. ANT needs a tensor dataset, which means we needed to convert the text files into a tensor dataset. PyTorch has a built-in function called ImageFolder, which allows the learner to load image files as tensor datasets.

For the conversion, the first step was encoding the image. Nominal value encoding was implemented for this project; then, we converted the encoded numbers into a colour format. For example, *add* operation code had an id of 1, which was converted into (0,0,1) for RGB colour. Then we used the PyPNG library to write the colour code into a PNG file [48]. Because tensor datasets need a quadratic image, and the operation code for one malware text file cannot form a quadratic image, padding it with (0,0,0) was applied to make this possible.

The benign and malicious folders were all generated with quadratic images. The ImageFolder function was then invoked to generate the tensor dataset. Each text file had a different length of operation code; therefore, the image size was different for each IoT program. However, ANT needs all data points to have the same size, so the built-in resize function was invoked with a parameter of 40 by 40. Then it was randomly split into three sub-datasets, namely, training dataset, validation dataset and test dataset. After the prepossessing, the three sub-tensor datasets were fed into ANT. Table 2 shows the configuration for the IoT dataset. Figure 3 explains it graphically. Also, the IoT Unseen dataset was provided and tested.

Table 2 IoT training configuration

Dataset	IoT
Batch-size	512
Epoch patience	5
Epoch node	100
Epoch finetune	200
Transformer	$2 \times$ Conv3-96 $+ 1 \times$ MaxPool
Router	$2 \times$ Conv3-48 $+$ GAP $+ 1 \times$ FC
Solver	Linear classifier
Batch normalization	Enabled
Max depth	10
Randomization seed	0

Fig. 3 Steps to get the IoT
dataset to work

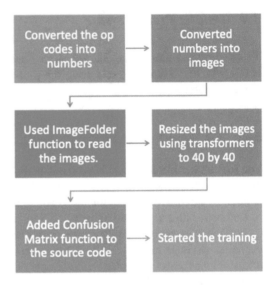

3.4.2 BATADAL and SWAT Datasets

BATADAL and SWAT datasets share some common attributes. Both are available as a NumPy array, which means that feature transformation is not necessary.

The first step was loading the NumPy array into the program. Because the range of numbers inside the dataset was extensive, we had to normalize them. The sklearn built-in normalization function was invoked to perform this exercise.

Like the IoT dataset, the number of features could not be morphed into a quadratic matrix, so it was padded with 0 to overcome this challenge. Because of the limitations of ANT, larger matrixes were needed in order to start the training, so Upsampling was applied to make this work. For BATADAL, the dataset was upsampled to 28 by 28, and for the SWAT dataset, it was upsampled to 54 by 54.

Also, the BATADAL and SWAT dataset did not have validation and test datasets; therefore, the built-in random split function was used to generate three sub-datasets: training, testing, and validation datasets. There is another issue the IoT dataset does not have, but BATADAL and SWAT have. It is the imbalance between normal and attack data points. Normal data points are exceedingly more than attack points; therefore, to prevent overfitting, the Imbalanced Dataset Sampler was utilized [49]. It solved the issue by oversampling low frequent classes and under-sampling high frequent ones.

After the prepossessing, the three sub-tensor datasets for both BATADAL and SWAT were fed into ANT. Table 3 shows the configuration for the BATADAL dataset, and Table 4 shows the training configuration for the SWAT dataset. Figure 4 below illustrates the process graphically.

Table 3 IoT training configuration based on SWAT datase

Dataset	BATADAL
Batch-size	256
Epoch patience	5
Epoch node	100
Epoch finetune	200
Transformer	$1 \times$ Conv3-96 + $1 \times$ MaxPool
Router	$1 \times$ Conv3-48 + GAP + Sigmoid
Solver	Linear classifier
Batch normalization	Enabled
Max depth	10
Randomization seed	0

Table 4 SWAT training configuration

Dataset	SWAT
Batch-size	256
Epoch patience	5
Epoch node	100
Epoch finetune	200
Transformer	$1 \times$ Conv3-5 + $1 \times$ MaxPool
Router	$1 \times$ Conv3-5 + GAP + $1 \times$ Sigmoid
Solver	Linear classifier
Downsample rate	1
Max depth	10
Randomization seed	1

Fig. 4 Steps to get the BATADAL and SWAT datasets to work

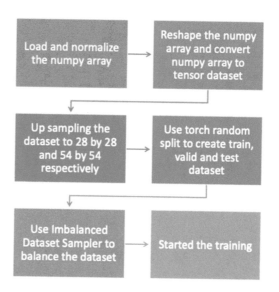

4   Results and Discussion

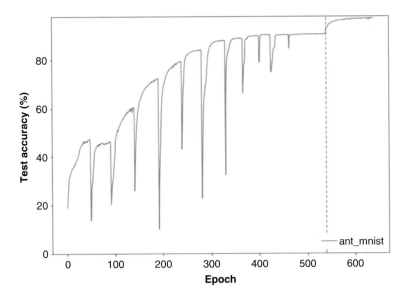

Fig. 5 Author's results

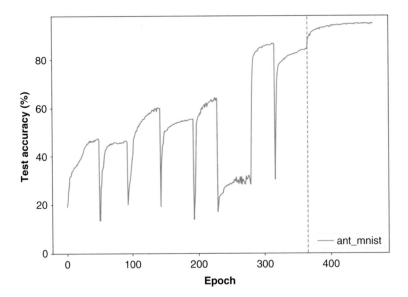

Fig. 6 Author's reproduced results

4.1 Abbreviations

TPR – True Positive Rate
FPR – False Positive Rate
TNR – True Negative Rate
FNR – False Negative Rate

4.2 Graphical Representations of the BATADAL, IoT and SWAT Dataset Results

This research consisted of two parts. Part one was to simulate the configuration environment used by the author and to determine if we could replicate his results using the same dataset. Figures 5 and 6 show that we were able to successfully duplicate the author's results using a similar configuration setting. The other part of the research was to conduct experiments with the other datasets provided to measure various attributes, as indicated in Table 5 above.

Table 5 also indicates that the BATADAL dataset produced the least accuracy percentage of 83.60%. That notwithstanding, the BATADAL dataset had a true-negative rate of 98.48%. The IoT and IoT unseen both produced accuracy rates of 98.04% and 99.45%; respectively, whiles the SWAT dataset delivered the highest accuracy of 100%.

Also, both the IoT and SWAT datasets correctly identified negatives at a rate of 100%. Precision for the IoT Unseen, IoT and the SWAT datasets was 1. Additionally, the true-positive rates for IoT, IoT Unseen and SWAT were 98.04%, 98.93% and 100%, respectively.

The false-positive rate and true-negative rate for the IoT unseen dataset is not applicable because the unseen dataset does not have negative samples; therefore, the FPR and TNR rates cannot be calculated. Figures 7, 8, and 9 show the graphical representations of the BATADAL, IoT and SWAT datasets, which support the numeral narratives in Table 5.

Table 5 Results for datasets

Dataset	Accuracy (%)	TPR (%)	FPR (%)	TNR (%)	FNR (%)	Precision	Recall (%)	F1 score
BATADAL	83.60	13.43	1.52	98.48	86.57	0.6528	13.43	0.2227
IoT Unseen	98.04	98.04	N/A	NA	1.96	1	98.04	0.9901
IoT	99.45	98.93	0	100	1.07	1	98.93	0.9946
SWAT	100	100	0	100	0	1	100	1

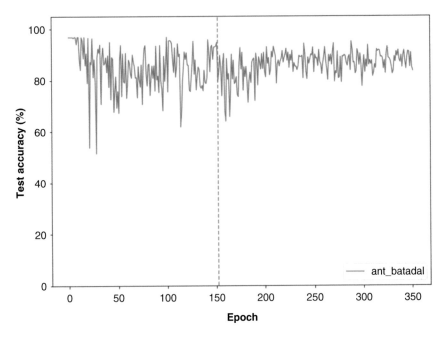

Fig. 7 BATADAL dataset result

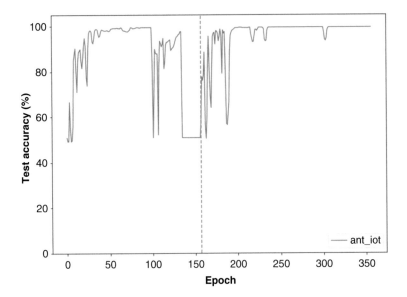

Fig. 8 IoT dataset result

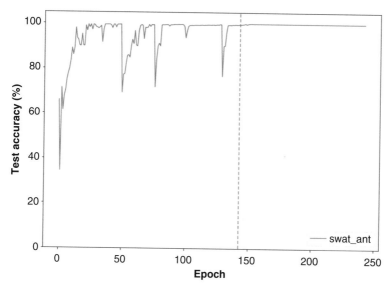

Fig. 9 SWAT dataset result

5 Conclusion and Future Work

In this report, ANT had been applied to the IoT, BATADAL and SWAT datasets with excellent results. Both IoT and SWAT datasets achieved an accuracy of 99.45% and 100%, respectively. For IoT Unseen dataset, an accuracy of 98.04% was achieved. Therefore, ANT would be suitable for some of the cybersecurity and threat intelligence tasks such as malware and network attack detection both from within and outside of an organization. As penetrating testing, malware and network attacks become more of a social engineering issue, the ability for companies to accurately predict and detect infiltration attacks from these sources will be a resource that will serve any company or government well, and we hope that this work will add to the knowledge base and advance learning in this domain.

Future work could apply the ANT to the Drone dataset and the Industrial Control System Cyberattack dataset. Tweaking the training parameters for BATADAL in order to get better results could also be an area of focus in the future. Also, BATADAL and SWAT have sub-datasets that contain many more data points that could be more extensively tested with ANT. Because of the RAM limitation, we could not load the whole dataset into our training system at this time. The IoT datasets do not have significant enough data points. Therefore, in future exercises, more IoT benign and malicious operation codes could be collected in order to improve the ANT model.

Acknowledgments Our sincere gratitude goes to Dr. Ali Dehghantanha, Ph.D., CISSP, CISM, Assistant Professor of the School of Computer Science and Director of the Cyber Science at the

University of Guelph and Hamed Haddadpajouh, Senior Researcher at the Cyber Science Lab. Without their assistance and direction, this whole research exercise would not have been possible.

References

1. A.N. Jahromi et al., An improved two-hidden-layer extreme learning machine for malware hunting. Comput. Secur. **89**, 101655 (2020)
2. A. Azmoodeh, A. Dehghantanha, K.-K.R. Choo, Big data and internet of things security and forensics: Challenges and opportunities, in *Handbook of Big Data and IoT Security*, (Springer, Cham, 2019), pp. 1–4. https://doi.org/10.1007/978-3-030-10543-3_1
3. H.M. Rouzbahani, H. Karimipour, A. Rahimnejad, A. Dehghantanha, G. Srivastava, Anomaly detection in cyber-physical systems using machine learning, in *Handbook of Big Data Privacy*, (Springer, Cham, 2020), pp. 219–235
4. H. Karimipour, A. Dehghantanha, R.M. Parizi, K.-K.R. Choo, H. Leung, A deep and scalable unsupervised machine learning system for cyber-attack detection in large-scale smart grids. IEEE Access **7**, 80778–80788 (2019)
5. N. Milosevic, A. Dehghantanha, K.-K.R. Choo, Machine learning aided android malware classification. Comput. Electr. Eng. **61**, 266–274 (2017)
6. W. Peters, A. Dehghantanha, R.M. Parizi, G. Srivastava, A comparison of state-of-the-art machine learning models for OpCode-based IoT malware detection, in *Handbook of Big Data Privacy*, (Springer, Cham, 2020), pp. 109–120
7. R. Raicea, Want to know how deep learning works? Here' s a quick guide for everyone (freeCodeCamp, 2017), https://www.freecodecamp.org/news/want-to-know-how-deep-learning-works-heres-a-quick-guide-for-everyone-1aedeca88076/. Accessed 20 Apr 2020
8. R. Talwar, A. Koury, Artificial intelligence–the next frontier in IT security? Netw. Secur. **2017**(4), 14–17 (2017)
9. A. Yazdinejad, R.M. Parizi, A. Dehghantanha, K.-K.R. Choo, P4-to-blockchain: A secure blockchain-enabled packet parser for software defined networking. Comput. Secur. **88** (2020). https://doi.org/10.1016/j.cose.2019.101629
10. A. Yazdinejad, A. Bohlooli, K. Jamshidi, Efficient design and hardware implementation of the OpenFlow v1.3 Switch on the Virtex-6 FPGA ML605. J. Supercomput. **74**(3) (2018). https://doi.org/10.1007/s11227-017-2175-7
11. Marketsandmarkets.com, Machine learning market by vertical (BFSI, healthcare and life sciences, retail, telecommunication, government and defense, manufacturing, energy and utilities), deployment mode, service, organization size, and region – Global For
12. BCC Research, Machine learning: Global markets to 2022 (2017)
13. A. Azmoodeh, A. Dehghantanha, K.-K.R. Choo, Robust malware detection for internet of (battlefield) things devices using deep eigenspace learning. IEEE Trans. Sustain. Comput. **4**(1), 88–95 (2018)
14. S. Nakhodchi, A. Dehghantanha, H. Karimipour, Privacy and security in smart and precision farming: A bibliometric analysis, in *Handbook of Big Data Privacy*, (Springer, Cham, 2020), pp. 305–318
15. S. Walker-Roberts, M. Hammoudeh, A. Dehghantanha, A systematic review of the availability and efficacy of countermeasures to internal threats in healthcare critical infrastructure. IEEE Access **6**, 25167–25177 (2018, March). https://doi.org/10.1109/ACCESS.2018.2817560
16. Technavio Research, Global machine learning market 2020–2024 | Increasing adoption of cloud-based offerings to boost the market growth | Technavio (Technavio Research, 2020), https://www.businesswire.com/news/home/20200327005177/en/Global-Machine-Learnin
17. S.M. Tahsien, H. Karimipour, P. Spachos, Machine learning based solutions for security of Internet of Things (IoT): A survey. J. Netw. Comput. Appl. **161**, 102630 (2020)

18. A. Yazdinejad, R.M. Parizi, G. Srivastava, A. Dehghantanha, K.-K.R. Choo, Energy efficient decentralized authentication in internet of underwater things using blockchain, in *2019 IEEE Globecom Workshops (GC Wkshps)*, (2019), pp. 1–6

19. A. Yazdinejad, H. HaddadPajouh, A. Dehghantanha, R.M. Parizi, G. Srivastava, M.-Y. Chen, Cryptocurrency malware hunting: A deep recurrent neural network approach. Appl. Soft Comput. J. Elsevier **96**, 106630 (2020)

20. M. Aledhari, R. Razzak, R.M. Parizi, F. Saeed, Federated learning: A survey on enabling technologies, protocols, and applications. IEEE Access **8**, 140699–140725 (2020). https://doi.org/10.1109/ACCESS.2020.3013541

21. A. Yazdinejad, R.M. Parizi, A. Dehghantanha, H. Karimipour, G. Srivastava, M. Aledhari, Enabling drones in the internet of things with decentralized blockchain-based security. IEEE Internet Things J., 1 (2020). https://doi.org/10.1109/jiot.2020.3015382

22. V. Mothukuri, R.M. Parizi, S. Pouriyeh, Y. Huang, A. Dehghantanha, G. Srivastava, A survey on security and privacy of federated learning. Futur. Gener. Comput. Syst. **115**, 619–640 (2020)

23. R.M. Parizi, S. Homayoun, A. Yazdinejad, A. Dehghantanha, K.-K.R. Choo, Integrating privacy enhancing techniques into blockchains using sidechains, in *IEEE Canadian Conference of Electrical and Computer Engineering, CCECE 2019*, (2019). https://doi.org/10.1109/CCECE.2019.8861821

24. A. Yazdinejad, R.M. Parizi, A. Dehghantanha, G. Srivastava, S. Mohan, A.M. Rababah, Cost optimization of secure routing with untrusted devices in software defined networking. J. Parallel Distrib. Comput. **143**, 36–46 (2020)

25. E.M. Dovom, A. Azmoodeh, A. Dehghantanha, D.E. Newton, R.M. Parizi, H. Karimipour, Fuzzy pattern tree for edge malware detection and categorization in IoT. J. Syst. Archit. **97**, 1–7 (2019)

26. H. Darabian et al., Detecting cryptomining malware: A deep learning approach for static and dynamic analysis. J. Grid Comput., 1–11 (2020)

27. A. Al-Abassi, H. Karimipour, A. Dehghantanha, R.M. Parizi, An ensemble deep learning-based cyber-attack detection in industrial control system. IEEE Access **8**, 83965–83973 (2020)

28. A. Azmoodeh, A. Dehghantanha, M. Conti, K.-K.R. Choo, Detecting crypto-ransomware in IoT networks based on energy consumption footprint. J. Ambient. Intell. Humaniz. Comput. **9**(4), 1141–1152 (2018)

29. P. Gupta, Decision trees in machine learning – Towards data science (2017), https://towardsdatascience.com/decision-trees-in-machinelearning-641b9c4e8052

30. Z. Wen, B. He, R. Kotagiri, S. Lu, J. Shi, Efficient gradient boosted decision tree training on GPUs, in *2018 IEEE International Parallel and Distributed Processing Symposium (IPDPS)*, (2018), pp. 234–243

31. H. Hashemi, A. Azmoodeh, A. Hamzeh, S. Hashemi, Graph embedding as a new approach for unknown malware detection. J. Comput. Virol. Hacking Technol. **13**(3), 153–166 (2017)

32. A. Azmoodeh, A. Dehghantanha, R.M. Parizi, S. Hashemi, B. Gharabaghi, G. Srivastava, Active spectral botnet detection based on eigenvalue weighting, in *Handbook of Big Data Privacy*, (Springer, Cham, 2020), pp. 385–397. https://doi.org/10.1007/978-3-030-38557-6_19

33. H.H. Pajouh, R. Javidan, R. Khayami, D. Ali, K.-K.R. Choo, A two-layer dimension reduction and two-tier classification model for anomaly-based intrusion detection in IoT backbone networks. IEEE Trans. Emerg. Top. Comput. **7**, 314–323 (2016)

34. S. Mohammadi, H. Mirvaziri, M. Ghazizadeh-Ahsaee, H. Karimipour, Cyber intrusion detection by combined feature selection algorithm. J. Inf. Secur. Appl. **44**, 80–88 (2019)

35. H. Liu, M. Cocea, W. Ding, Decision tree learning based feature evaluation and selection for image classification, in *2017 International Conference on Machine Learning and Cybernetics (ICMLC)*, vol. 2, (2017), pp. 569–574

36. H. Hosseini, B. Xiao, M. Jaiswal, R. Poovendran, On the limitation of convolutional neural networks in recognizing negative images, in *2017 16th IEEE International Conference on Machine Learning and Applications (ICMLA)*, (2017), pp. 352–358

37. A. Yazdinejad, G. Srivastava, R.M. Parizi, A. Dehghantanha, H. Karimipour, S.R. Karizno, SLPoW: Secure and low latency proof of work protocol for blockchain in green IoT networks, in *2020 IEEE 91st Vehicular Technology Conference (VTC2020-Spring)*, (2020), pp. 1–5
38. A. Yazdinejad, R.M. Parizi, A. Dehghantanha, K.-K.R. Choo, Blockchain-enabled authentication handover with efficient privacy protection in SDN-based 5G networks. IEEE Trans. Netw. Sci. Eng. (2019). https://doi.org/10.1109/TNSE.2019.2937481
39. A. Singh, K. Click, R.M. Parizi, Q. Zhang, A. Dehghantanha, K.-K.R. Choo, Sidechain technologies in blockchain networks: An examination and state-of-the-art review. J. Netw. Comput. Appl. **149**, 102471 (2020). https://doi.org/10.1016/j.jnca.2019.102471
40. A. Yazdinejad, R.M. Parizi, A. Dehghantanha, Q. Zhang, K.-K.R. Choo, An energy-efficient SDN controller architecture for IoT networks with blockchain-based security. IEEE Trans. Serv. Comput. (2020). https://doi.org/10.1109/TSC.2020.2966970
41. D. Połap, G. Srivastava, A. Jolfaei, R.M. Parizi, Blockchain technology and neural networks for the internet of medical things, in *IEEE INFOCOM 2020 – IEEE Conference on Computer Communications Workshops (INFOCOM WKSHPS)*, (2020), pp. 508–513. https://doi.org/10.1109/INFOCOMWKSHPS50562.2020.9162735
42. A. Yazdinejad, G. Srivastava, R.M. Parizi, A. Dehghantanha, K.-K.R. Choo, M. Aledhari, Decentralized authentication of distributed patients in hospital networks using blockchain. IEEE J. Biomed. Health Inform. **24**(8), 2146–2156 (2020)
43. Q. Chen, G. Srivastava, R.M. Parizi, M. Aloqaily, I. Al Ridhawi, An incentive-aware blockchain-based solution for internet of fake media things. Inf. Process. Manag., 102370 (2020). https://doi.org/10.1016/j.ipm.2020.102370
44. A. Yazdinejad, R.M. Parizi, A. Bohlooli, A. Dehghantanha, K.-K.R. Choo, A high-performance framework for a network programmable packet processor using P4 and FPGA. J. Netw. Comput. Appl. **156**, 102564 (2020)
45. H. Darabian et al., A multiview learning method for malware threat hunting: Windows, IoT and android as case studies. World Wide Web **23**(2), 1241–1260 (2020)
46. H. HaddadPajouh, A. Dehghantanha, R. Khayami, K.-K.R. Choo, A deep recurrent neural network based approach for Internet of Things malware threat hunting. Futur. Gener. Comput. Syst. **85**, 88–96 (2018). https://doi.org/10.1016/j.future.2018.03.007
47. R. Tanno, K. Arulkumaran, D.C. Alexander, A. Criminisi, A. Nori, Adaptive neural trees, in *36th International Conference on Machine Learning, ICML 2019*, vol. 2019-June, (2019, July), pp. 10761–10770
48. Drj11, PyPNG documentation – PyPNG 0.0.17 documentation (Github, 2019), https://pypng.readthedocs.io/en/latest/index.html. Accessed 20 Apr 2020
49. Ufoym, GitHub – ufoym/imbalanced-dataset-sampler: A (PyTorch) imbalanced dataset sampler for oversampling low frequent classes and undersampling high frequent ones (GitHub, 2020), https://github.com/ufoym/imbalanced-dataset-sampler. Accessed 20 Apr 2020

Evaluating Performance of Scalable Fair Clustering Machine Learning Techniques in Detecting Cyber Attacks in Industrial Control Systems

Akansha Handa and Prabhat Semwal

1 Introduction

The Internet of things (IoT) devices are already being used to make important decisions in our society [1–7]. For example, managing critical Industrial Control Systems (ICS) like nuclear plants and other huge infrastructure with IoT devices to crime investigation using facial recognition application [8–12]. Thus, prominence of machine learning algorithms in such IoT and ICS ecosystems as a background technology has led to growing concern regarding the biased decisions being made by these algorithms [13–15]. Although, it seems if machine learning algorithms will be used to make decisions instead of humans, the outcome will be unbiased. However, these algorithms can be biased if the underlying training data has any sort of discriminating factor in it. Therefore, built machine learning models can have biased behavior and can raise multiple security concerns related to IoT devices [16–18]. As a result, a significant amount of research has been performed to understand the fairness factor in machine learning algorithms [13, 19–21].

Clustering is a fundamental unsupervised machine learning technique which divides the data points into more related groups (clusters) without the knowledge of the label of the data points which is like data analysis and data mining processes [22–26]. As a result, Clustering has been extensively studied from a fair machine learning perspective. The first fair unsupervised learning algorithms was explored in the form of fair clustering which introduced the concept of *fairlet decomposition* i.e. is to partition data points into a small cluster (*fairlets*) with the constraint of maintaining balance in terms of defined sensitive features, such as race or gender, and cluster those balanced fairlets into overall balanced clusters [27]. To measure

A. Handa (✉) · P. Semwal
School of Computer Science, University of Guelph, Guelph, ON, Canada
e-mail: ahanda@uoguelph.ca; psemwal@uoguelph.ca

© Springer Nature Switzerland AG 2021
K.-K. R. Choo, A. Dehghantanha (eds.), *Handbook of Big Data Analytics and Forensics*, https://doi.org/10.1007/978-3-030-74753-4_7

the fairness in clustering, they defined the balance as a ratio of the number of two colors in each cluster. Hence, defined the balance of one cluster(S) to be min $\left(\frac{Sr}{Sb}\frac{Sb}{Sr}\right)$ where Sr and Sb are the subsets of red and blue points in S and the overall balance of clustering to be a minimum balance of any one cluster. The proposed fairlet decomposition algorithms are suitable for two standard clustering objectives: K-center and K-median.

While the fairlet decomposition solution introduced fairness concept in clustering, it lacked in handling the large datasets due to quadratic running time issue of the algorithm on performing fair let decomposition. The scalable fair clustering provides the solution to this limitation as an approximation algorithm to compute fairlet decomposition cost with running time in near-linear, where cost represents the overall distance between data points and their centers at fairlet/clusters [28]. The fairlet decomposition algorithms introduced in [28], focused only on the K-median clustering objective to avoid the issue of sensitive attributes outliers caused by K-center.

In this paper, we use the scalable fair clustering approach to study the nearly linear running time result of proposed fairlet decomposition (FD) algorithm. We have defined two objectives of our experiment: first, to calculate the running time of the FD model on the three different datasets than base paper: IOT [29, 30] ICS and SWAT. Second, to evaluate the performance of FD model on the base of five widely used evaluation metrics Accuracy, FPR, TPR, Precision and F1-Score, described in Sects. 4.1 and 4.2.

The Sect. 2 of this paper presents a literature review on related academic research papers from recent years. Section 3 describes the methodology used in this work, Sect. 4 shows the result of our study and in Sect. 5 we draw a conclusion of our experiment and suggest future work. The references are listed in the last section.

2   Related Works

Over the last few years, several studies have been presented exploring the IoT and ICS systems from the security perspective [31–38] identified most common cybersecurity threats and incidents patterns using VCDB cybersecurity dataset and based on their findings indicated the need for reassessment of cybersecurity standards. Sakhnini et al. [39] conducted a comprehensive survey of all research articles and papers on the security aspect of the IoT device, such as the Smart grid. In this survey, they illustrate the range of cyber-threats posed by a Smart grid and future research aspect of smart grid security.

While various studies examined the security aspects of IoT and ICS systems, others have proposed ways to detect cyber-attacks on these systems using machine learning techniques [40–43]. Dovom et al. [44] performed the classification and detection of malware in IoT using fuzzy and fuzzy patter (FPT) machine learning techniques. They used operational codes of malware and benign samples of IoT systems and established FPT can classify malware with the accuracy of 99.8%. Similarly, [45] used the operational codes-based technique and different machine learning techniques to detect IoT malware. There is well-established research has been presented to detect malware or other cyber-attacks detection using machine learning. However, such work does not investigate the accuracy of the machine learning outcomes in terms of fairness notion.

In recent years, researchers have noted a need to study the fairness notion in machine learning techniques and developed different types of fair machine learning models based on both supervised and unsupervised machine learning [45, 46]. Lee et al. [47] proposed the fairness notion by performing fair feature subset selection algorithm capable of performing high-speed and high-accuracy classification. Grari et al. [48] established a novel approach of gradient tree boosting with the capability of maintaining the same level of fairness with a high accuracy rate than other Fair classifier models. Some proposed to deal with the discriminating factor by removing the biased data from datasets which are used to train such machine learning models. Veale and Binns [49] proposed multiple ways to improve the fairness factor of machine learning by addressing discriminating data such as sensitive attributes.

As supervised, the fairness notion has also been addressed in unsupervised machine learning [26]. Most of the fair clustering studies have extended the principle of fairness developed by [27]. Bera et al. [50] proposed a more generalized fair clustering model that can transform any classical algorithms (k-center, k-median, k-means) into fair algorithms with a slight decrease in the overall quality. The developed algorithm defined l (number of protected features) >= 2 and allowed overlapping of features to cover data points like (African-America). Also, the improved K-median clustering cost was achieved for all three datasets (census, bank, diabetes) than reported in [27, 28]. Abraham and Sundaram [51] proposed a novel clustering solution, FairKM, that scaled the existing clustering solution for k-means. They concluded the outcome of FairKM in terms of both quality and fairness of clustering. Huang et al. [52] and Schmidt et al. [53] considered the problem of scaling the existing fair clustering solutions for large datasets with the help of corset solution to reduce the input data size.

In [54], fair clustering solution is scaled up to deal with multiple colors (i.e. $l > =3$). Also, provided no-constant factor clustering (k-clustering) that is suitable for any center-based k-clustering objective (K-means, K-median, K-center). Similarly, [55] proposed a scalable approach by handling sensitive or protected attributes for large class clustering objectives: K-center, K-means, K-median. On the contrary [56], introduced fairness in terms of proportionality of by replacing the notion of the sensitive attribute to make fair clusters.

The above research work incorporates the fairness notion in machine learning but does not measure the effectiveness of fair machine learning models. The evaluation of the machine learning models in terms of accuracy is necessary and vital to use advanced machine learning techniques in critical systems like IoT and ICS. This motivates our work to execute the Scalable fair clustering machine learning technique and evaluate the performance of the fair machine learning model in terms of both quality and fairness.

3   Methodology

In this section, we illustrate the experimental study carried out to determine the time linearity result and performance evaluation of scalable fair clustering algorithm for three datasets: IoT, CIS and SWAT. First, we describe the process followed to process all three datasets to extract the optimum features required for the FD algorithm implementation and then explain the design of the FD model for the fairlet decomposition cost runtime calculation part. Finally, we describe the steps taken to integrate the performance assessment component into the FD model to calculate the evaluation metrics.

3.1   IoT Dataset

The IoT dataset represents the network traffic of IoT devices. It contains samples in the form of benign IoT traffic and traffic where malware has been executed on IoT devices.

3.1.1   Dataset Processing & Feature Extraction

The provided dataset consisted of samples in the form of opcode files, in which each file is either a benign sample text file or malware sample text file. Thus, each file is processed to transform the data in text format (opcodes) into numerical data, using the Term Frequency – Inverse Document Frequency (TF-IDF) technique. The TD-IDF score of each benign and malware sample was used as a feature and the TD-IDF scores for those obtained features were used as feature values. As a result, a total of 236 features and 512 samples were extracted, where total samples contained 244 malware and 258 benign class samples.

To reduce the dimensionality of the processed dataset, feature selection was performed using Univariate feature selection and Tree-based feature selection method: ExtraTreeClassifier. Total five features were selected based on the highest-ranking feature and were more common in both feature selection output. The five features were extracted into a separate file to create the low dimensional dataset, listed in Table 1. We will refer the dataset as "IOT" dataset.

Table 1 Description of three datasets used for our experiment

Dataset	Dimension	Number of points	Sensitive attribute
IOT	4	512	V227
ICS	5	11,439	S_PU8
SWAT	5	14,994	P3_STATE

3.2   ICS Dataset

The industrial control systems are the cyber-physical systems, such as water distribution systems, soil treatment plants, and other geographically distributed systems controlled by computers. The ICS dataset contains the cyber-attack data of such control systems. Many ICS datasets are already being used to analyse the security threat to any ICS [57]. It contains the data form Intrusion detection systems, sensors, network, SCADA operations.

3.2.1   Dataset Processing & Feature Extraction

The provided ICS dataset consists of 22 features and a total of 11,439 samples, where overall samples were comprised of 492 attack and 10,947 normal class samples. The dataset was processed to extract the best features for the FD model using a feature selection method: ExtraTreeClassifier. The five most common and top-ranked features were extracted into a csv file, in which four features were protected feature and one feature was a sensitive, mentioned in Table 1. We will refer the processed dataset as "ICS" dataset.

3.3   SWAT Dataset

The Secure Water Treatment Plant (SWAT) is a cyber-physical system. The SWAT dataset is the data collected from the water treatment plant on both normal operating days and cyber-attack days. This dataset consists of data from control systems like sensors, alarms, water pumps [58].

3.3.1   Dataset Processing & Feature Extraction

The provided dataset consists of 77 features and 14,994 samples, with 5474 attack and 9521 normal samples. We processed the dataset to reduce the dimensionality of the dataset to a smaller number of protected features using Univariate feature selection and ExtraTreeClassifier method. The highest ranked and most common features were extracted by comparing the results of two features selection methods. We will refer the processed SWAT dataset as "SWAT-P" dataset. The final layout of "SWAT" dataset is described in Table 1.

3.4   Fairlet Decomposition Model

The FD model used in our experiment is built by using the approximate fairlet decomposition algorithm version of scalable fair clustering, for which practical implementation is provided in [28]. Our FD model is designed to achieve the objective of our experiment, mentioned in the section, on the form of two components: Fairlet decomposition cost and K-medoid.

3.4.1   Fairlet Decomposition Cost

To accomplish the time-linearity objective for all three processed datasets, as mentioned in Sect. 1, we have used the fairlet decomposition cost part of FD algorithm. We trained our FD model on a processed dataset and determined the optimal values of r and b (i.e. red and blue), through trial and error, for the given dataset. In our experiment, we achieved the same balance b = 0.5 for all three processed datasets. Second, we divided each one of the processed datasets into multiple sub-samples. As we achieved the same balance (0.5) for all three datasets, we have executed the FD model on all the sub-samples of a processed dataset with the same balance. As a result, the run time of the FD model was recorder for performing fairlet decomposition cost for each sub-sample of a given dataset.

3.4.2   K-medoids

To achieve the second objective of our study, mentioned in Sect. 1, we have used the K-medoid part of the FD algorithm, where K-medoid is used to cluster the (r, b) balanced fairlets with the objective of minimizing the overall distance between the datapoints in a cluster and points which are fairlets/clusters center (centroid or medoids) [28, 59]. We used the K-medoid outcome the IDX vector (or cluster indexes of each observation) to calculate the predicted label of a given dataset, where the IDX value of each sample represents the cluster number in which a sample is clustered. Therefore, we mapped the samples IDX values with their respective actual class labels in order to label the IDX (cluster) as either of two class label (like 1 = benign and 0 = malware sample class) by determining the which type of class samples have maximum weightage in a formed cluster (IDX).

For example, with cluster k = 2, the IDX (medoid) consisted of value 1 and 2 for each sample, indicating the predicted group or cluster formed on running FD model on a processed dataset, such as IOT dataset. The formed clusters (IDX) were labelled as benign if the maximum number of samples with IDX = 1 has actual class label = 0 (benign) and the second cluster (IDX = 2) was labelled as malware. As a result, created the predicted label by marking all the samples with IDX = 1 as being and IDX = 2 as malware. Also, to evaluate the performance of FD model for all three datasets, we have performed similar IDX mapping to compute the accuracy and other evaluation metrics.

4 Results & Discussion

This section highlights the results achieved with the fair clustering method on three processed datasets, listed in Table 1. First, we explain the results achieved for time linearity objective. Second, we describe the evaluation metrics used to measure the performance of FD model; then we list the results achieved with the FD model for the defined evaluation metrics. Finally, we perform the performance analysis of FD model using the results achieved for all three processed datasets.

4.1 Fairlet Decomposition Cost Result

In our experiment, we have calculated the time linearity for all three datasets by dividing each dataset into sub-samples and running our FD model on theses sub-samples. We calculated the run time of FD model for fairlet decomposition cost with the optimal balance of $b = 0.5$ (i.e. $r = 1$ and $b = 2$) and number of clusters $k = 20$ for all three datasets.

As shown in Fig. 1, the observed run time of FD model was plotted against the number of samples and running time in seconds. For all three datasets, the FD model performed the fair clustering at in nearly linear time format.

4.2 Evaluation Measures

In our experiment, we evaluate the performance of FD model by creating the confusion matrix with the actual and generated predicted labels of FD model. The most common metrics: Accuracy, TPR, FPR, Precision and F1-score were calculated by relating the predicted outcome in the form of confusion matrix: True Positive (TP), True Negative (TN), False Positive (FP), False Negative (FN). Table 2 contains the description all the evaluation measure used for computing the performance of FD model.

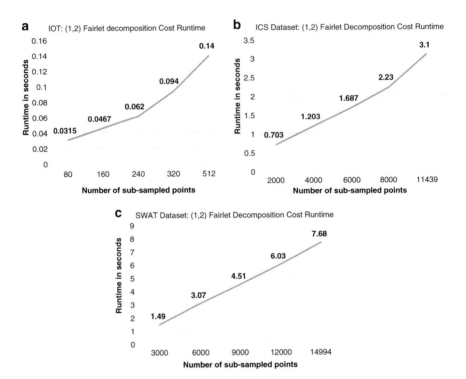

Fig. 1 Each figure represents the FD model running time in seconds for defined balanced and number of clusters for datasets: (**a**) IOT (**b**) ICS and (**c**) SWAT

Table 2 Evaluation metrics used for FD model performance analysis

Evaluation metric	Formula
Accuracy	$\frac{TP+TN}{TP+TN+FP+FN}$
TPR/recall	$\frac{TP}{TP+FN}$
FPR	$\frac{FP}{FP+TN}$
Precision	$\frac{TP}{TP+FP}$
F1-score	$2 * \frac{(precision*recall)}{(precision+recall)}$

4.3 FD Model Results

On running the FD model on all three datasets, with balance of 0.5, we computed the predicted label, as described in Sect. 3.2. The accuracy achieved with IOT, ICS and SWAT dataset is 83.48%, 95.82%, and 88.59% respectively.

Table 3 Describes the overall results achieved with FD model for IOT, ICS and SWAT datasets

Dataset	Clusters (K)	Accuracy	TPR	FPR	Precision	F1-score
IOT	15	83.48	77.19	10.0	88.8	82.65
ICS	10	95.82	98.92	68.57	96.78	97.83
SWAT	12	88.59	84	58.12	88.59	93.95

4.4 Result Analysis

As shown in Table 3, we evaluate the FD model based on evaluation metrics calculated for IOT, ICS and SWAT dataset. The overall performance of FD model is evaluated in terms of accuracy of model in clustering samples as per the defined sensitive attribute of a given dataset. We achieved the high accuracy of fair clustering with ICS dataset and SWAT, 95.82% and 88.59% respectively. However, the falsely reported samples for these two datasets in terms of FPR, 68% with ICS dataset and 58.12% with SWAT dataset, are higher than IOT dataset with FPR of 10%. The major reason for this was the data dependency of Fairlet Decomposition Model. The ICS dataset with 492 attack and 10,947 normal samples and SWAT with 5474 attack and 9521 normal samples are highly unbalanced in terms of two classes due to which the clustering performed by the model is somehow biased. On the other hand, IOT dataset has balanced classification of target label (244 malware and 258 benign) therefore FD model evaluation of IOT dataset is considered more accurate with an overall accuracy of 83.48% and a low FPR of 10%.

5   Conclusion & Future Work

In this paper, we studied about the scalable fair clustering algorithm and implemented Fairlet Decomposition Model to evaluate the overall performance of the fair clustering approach. We evaluated the performance of FD model in terms of handling the range of datasets while maintaining run time linearity and accuracy achieved by the FD model on three different datasets. We have observed a tradeoff based on data provided for the evaluation. There are several challenges like defining an appropriate version of fairness. An immediate future direction is to enhance the performance of Fairlet Decomposition Model in term of dataset independency.

References

1. M. Conti, A. Dehghantanha, K. Franke, S. Watson, Internet of Things security and forensics: Challenges and opportunities. Futur. Gener. Comput. Syst. **78**, 544–546 (2018). https://doi.org/10.1016/j.future.2017.07.060

2. H. HaddadPajouh, R. Khayami, A. Dehghantanha, K.-K.R. Choo, R.M. Parizi, AI4SAFE-IoT: An AI-powered secure architecture for edge layer of Internet of things. Neural Comput. Applic. **32**(20), 16119–16133 (2020). https://doi.org/10.1007/s00521-020-04772-3

3. S. Nakhodchi, A. Dehghantanha, H. Karimipour, Privacy and security in smart and precision farming: A bibliometric analysis, in *Handbook of Big Data Privacy*, (Springer, Cham, 2020), pp. 305–318

4. S. Walker-Roberts, M. Hammoudeh, A. Dehghantanha, A systematic review of the availability and efficacy of countermeasures to internal threats in healthcare critical infrastructure. IEEE Access **6**, 25167–25177 (2018, March). https://doi.org/10.1109/ACCESS.2018.2817560

5. A. Yazdinejad, R.M. Parizi, A. Dehghantanha, H. Karimipour, G. Srivastava, M. Aledhari, Enabling drones in the internet of things with decentralized blockchain-based security. IEEE Internet Things J., 1 (2020). https://doi.org/10.1109/jiot.2020.3015382

6. F. Daryabar, A. Dehghantanha, N.I. Udzir, S.B. Shamsuddin, Towards secure model for SCADA systems, in *Proceedings Title: 2012 International Conference on Cyber Security, Cyber Warfare and Digital Forensic (CyberSec)*, (2012), pp. 60–64

7. A. Yazdinejad, R.M. Parizi, A. Dehghantanha, K.-K.R. Choo, Blockchain-enabled authentication handover with efficient privacy protection in SDN-based 5G networks. IEEE Trans. Netw. Sci. Eng. (2019). https://doi.org/10.1109/TNSE.2019.2937481

8. D.C. Parkes, R.V. Vohra, Algorithmic and economic perspectives on fairness. arXiv Prepr. arXiv1909.05282 (2019)

9. M.M. Ahmadian, M. Shajari, M.A. Shafiee, Industrial control system security taxonomic framework with application to a comprehensive incidents survey. Int. J. Crit. Infrastruct. Prot. **29**, 100356 (2020)

10. B.J. Santos, R.P. Tabacow, M. Barboza, T.F. Leão, E.G.P. Bock, Cyber security in health: Standard protocols for IoT and supervisory control systems, in *Cyber Security of Industrial Control Systems in the Future Internet Environment*, (IGI Global, 2020), pp. 313–329

11. S.A.I. Shouborno et al., Complete automation of an E-commerce system with internet of things, in *2019 IEEE International Conference on Robotics, Automation, Artificial-Intelligence and Internet-of-Things (RAAICON)*, pp. 81–86

12. K. Bolouri, A. Azmoodeh, A. Dehghantanha, M. Firouzmand, Internet of things camera identification algorithm based on sensor pattern noise using color filter array and wavelet transform, in *Handbook of Big Data and IoT Security*, (Springer, Cham, 2019), pp. 211–223. https://doi.org/10.1007/978-3-030-10543-3_9

13. N. Mehrabi, F. Morstatter, N. Saxena, K. Lerman, A. Galstyan, A survey on bias and fairness in machine learning. arXiv Prepr. arXiv1908.09635 (2019)

14. A. Yazdinejad, A. Bohlooli, K. Jamshidi, Performance improvement and hardware implementation of Open Flow switch using FPGA, in *2019 5th Conference on Knowledge Based Engineering and Innovation (KBEI)*, (2019). https://doi.org/10.1109/KBEI.2019.8734914

15. A. Yazdinejad, A. Bohlooli, K. Jamshidi, Efficient design and hardware implementation of the OpenFlow v1.3 Switch on the Virtex-6 FPGA ML605. J. Supercomput. **74**(3) (2018). https://doi.org/10.1007/s11227-017-2175-7

16. H. HaddadPajouh, A. Dehghantanha, R.M. Parizi, M. Aledhari, H. Karimipour, A survey on internet of things security: Requirements, challenges, and solutions. Internet of Things **3**, 100129 (2019)

17. A. Dehghantanha, K.-K. R. Choo (eds.), *Handbook of Big Data and IoT Security* (Springer, Cham, 2019)

18. A. Azmoodeh, A. Dehghantanha, K.-K.R. Choo, Big data and internet of things security and forensics: Challenges and opportunities, in *Handbook of Big Data and IoT Security*, (Springer, Cham, 2019), pp. 1–4. https://doi.org/10.1007/978-3-030-10543-3_1

19. C. Dwork, M. Hardt, T. Pitassi, O. Reingold, R. Zemel, Fairness through awareness, in *Proceedings of the 3rd Innovations in Theoretical Computer Science Conference*, (2012), pp. 214–226

20. D. Slack, S.A. Friedler, E. Givental, Fairness warnings and fair-MAML: Learning fairly with minimal data, in *Proceedings of the 2020 Conference on Fairness, Accountability, and Transparency*, (2020), pp. 200–209

21. Fair algorithms for machine learning, in *Proceedings of the 2017 ACM Conference on Economics and Computation*, https://dl.acm.org/doi/abs/10.1145/3033274.3084096. Accessed 18 Sep 2020
22. M.Z. Rodriguez et al., Clustering algorithms: A comparative approach. PLoS One **14**(1), e0210236 (2019)
23. L. Rokach, O. Maimon, Clustering methods, in *Data Mining and Knowledge Discovery Handbook*, (Springer, New York, 2005), pp. 321–352
24. M. Ghesmoune, M. Lebbah, H. Azzag, State-of-the-art on clustering data streams. Big Data Anal. **1**(1), 13 (2016)
25. P. Berkhin, A survey of clustering data mining techniques, in *Grouping Multidimensional Data*, (Springer, Berlin, 2006), pp. 25–71
26. A. Azmoodeh, A. Dehghantanha, R.M. Parizi, S. Hashemi, B. Gharabaghi, G. Srivastava, Active spectral botnet detection based on eigenvalue weighting, in *Handbook of Big Data Privacy*, (Springer, Cham, 2020), pp. 385–397. https://doi.org/10.1007/978-3-030-38557-6_19
27. F. Chierichetti, R. Kumar, S. Lattanzi, S. Vassilvitskii, Fair clustering through fairlets, in *Advances in Neural Information Processing Systems*, (MIT Press, Cambridge, MA, 2017), pp. 5029–5037
28. A. Backurs, P. Indyk, K. Onak, B. Schieber, A. Vakilian, T. Wagner, Scalable fair clustering. arXiv Prepr. arXiv1902.03519 (2019)
29. A. Azmoodeh, A. Dehghantanha, K.-K.R. Choo, Robust malware detection for internet of (battlefield) things devices using deep eigenspace learning. IEEE Trans. Sustain. Comput. **4**(1), 88–95 (2018)
30. H. HaddadPajouh, A. Dehghantanha, R. Khayami, K.-K.R. Choo, A deep recurrent neural network based approach for internet of things malware threat hunting. Futur. Gener. Comput. Syst. **85**, 88–96 (2018). https://doi.org/10.1016/j.future.2018.03.007
31. A. Yazdinejad, R.M. Parizi, A. Bohlooli, A. Dehghantanha, K.-K.R. Choo, A high-performance framework for a network programmable packet processor using P4 and FPGA. J. Netw. Comput. Appl. **156**, 102564 (2020)
32. Q. Chen, G. Srivastava, R.M. Parizi, M. Aloqaily, I. Al Ridhawi, An incentive-aware blockchain-based solution for internet of fake media things. Inf. Process. Manag., 102370 (2020). https://doi.org/10.1016/j.ipm.2020.102370
33. A. Yazdinejad, R.M. Parizi, G. Srivastava, A. Dehghantanha, K.-K.R. Choo, Energy efficient decentralized authentication in internet of underwater things using blockchain, in *2019 IEEE Globecom Workshops (GC Wkshps)*, (2019), pp. 1–6
34. A. Yazdinejad, H. HaddadPajouh, A. Dehghantanha, R.M. Parizi, G. Srivastava, M.-Y. Chen, Cryptocurrency malware hunting: A deep recurrent neural network approach. Appl. Soft Comput. Elsevier **96**, 106630 (2020)
35. M. Aledhari, R. Razzak, R.M. Parizi, F. Saeed, Federated learning: A survey on enabling technologies, protocols, and applications. IEEE Access **8**, 140699–140725 (2020). https://doi.org/10.1109/ACCESS.2020.3013541
36. A. Yazdinejad, R.M. Parizi, A. Dehghantanha, Q. Zhang, K.-K.R. Choo, An energy-efficient SDN controller architecture for IoT networks with blockchain-based security. IEEE Trans. Serv. Comput. **13**(4), 625–638 (2020)
37. S. Walker-Roberts, M. Hammoudeh, O. Aldabbas, M. Aydin, A. Dehghantanha, Threats on the horizon: Understanding security threats in the era of cyber-physical systems. J. Supercomput. **76**(4), 2643–2664 (2020)
38. A. Yazdinejad, R.M. Parizi, A. Dehghantanha, G. Srivastava, S. Mohan, A.M. Rababah, Cost optimization of secure routing with untrusted devices in software defined networking. J. Parallel Distrib. Comput. **143**, 36–46 (2020)
39. J. Sakhnini, H. Karimipour, A. Dehghantanha, R.M. Parizi, G. Srivastava, Security aspects of Internet of Things aided smart grids: A bibliometric survey. Internet of things, 100111 (2019)

40. A. Azmoodeh, A. Dehghantanha, M. Conti, K.-K.R. Choo, Detecting crypto-ransomware in IoT networks based on energy consumption footprint. J. Ambient. Intell. Humaniz. Comput. **9**(4), 1141–1152 (2018)
41. M. Saharkhizan, A. Azmoodeh, A. Dehghantanha, K.-K.R. Choo, R.M. Parizi, An ensemble of deep recurrent neural networks for detecting IoT cyber attacks using network traffic. IEEE Internet Things J. **7**(9), 8852–8859 (2020). https://doi.org/10.1109/jiot.2020.2996425
42. H. Darabian et al., A multiview learning method for malware threat hunting: Windows, IoT and android as case studies. World Wide Web **23**(2), 1241–1260 (2020)
43. H. Darabian, A. Dehghantanha, S. Hashemi, S. Homayoun, K.R. Choo, An opcode-based technique for polymorphic Internet of Things malware detection. Concurr. Comput. Pract. Exp. **32**(6), e5173 (2020)
44. E.M. Dovom, A. Azmoodeh, A. Dehghantanha, D.E. Newton, R.M. Parizi, H. Karimipour, Fuzzy pattern tree for edge malware detection and categorization in IoT. J. Syst. Archit. **97**, 1–7 (2019)
45. F. Kamiran, T. Calders, Classifying without discriminating, in *2009 2nd International Conference on Computer, Control and Communication*, (2009), pp. 1–6
46. A. Pérez-Suay, V. Laparra, G. Mateo-García, J. Muñoz-Marí, L. Gómez-Chova, G. Camps-Valls, Fair kernel learning, in *Joint European Conference on Machine Learning and Knowledge Discovery in Databases*, (2017), pp. 339–355
47. H.-M. Lee, C.-M. Chen, C.-C. Tan, An intelligent web-page classifier with fair feature-subset selection, in *Proceedings Joint 9th IFSA World Congress and 20th NAFIPS International Conference (Cat. No. 01TH8569)*, vol. 1, (2001), pp. 395–400
48. V. Grari, B. Ruf, S. Lamprier, M. Detyniecki, Fair adversarial gradient tree boosting, in *2019 IEEE International Conference on Data Mining (ICDM)*, (2019), pp. 1060–1065
49. M. Veale, R. Binns, Fairer machine learning in the real world: Mitigating discrimination without collecting sensitive data. Big Data Soc. **4**(2), 2053951717743530 (2017)
50. S. Bera, D. Chakrabarty, N. Flores, M. Negahbani, Fair algorithms for clustering, in *Advances in Neural Information Processing Systems*, (MIT Press, Cambridge, MA, 2019), pp. 4954–4965
51. S.S. Abraham, S.S. Sundaram, Fairness in clustering with multiple sensitive attributes. arXiv Prepr. arXiv1910.05113 (2019)
52. L. Huang, S. Jiang, N. Vishnoi, Coresets for clustering with fairness constraints, in *Advances in Neural Information Processing Systems*, (MIT Press, Cambridge, MA, 2019), pp. 7589–7600
53. M. Schmidt, C. Schwiegelshohn, C. Sohler, Fair coresets and streaming algorithms for fair k-means clustering. arXiv Prepr. arXiv1812.10854 (2018)
54. M. Böhm, A. Fazzone, S. Leonardi, C. Schwiegelshohn, Fair clustering with multiple colors. arXiv Prepr. arXiv2002.07892 (2020)
55. I.M.Z.E.G. Jing, Y.I.B. Ayed, Clustering with fairness constraints: A flexible and scalable approach. CoRR (2019)
56. X. Chen, B. Fain, C. Lyu, K. Munagala, Proportionally fair clustering. arXiv Prepr. arXiv1905.03674 (2019)
57. S. Choi, J.-H. Yun, S.-K. Kim, A comparison of ICS datasets for security research based on attack paths, in *International Conference on Critical Information Infrastructures Security*, (2018), pp. 154–166
58. J. Goh, S. Adepu, K.N. Junejo, A. Mathur, A dataset to support research in the design of secure water treatment systems, in *International Conference on Critical Information Infrastructures Security*, (2016), pp. 88–99
59. T.S. Madhulatha, Comparison between k-means and k-medoids clustering algorithms, in *International Conference on Advances in Computing and Information Technology*, (2011), pp. 472–481

Fuzzy Bayesian Learning for Cyber Threat Hunting in Industrial Control Systems

Kassidy Marsh and Samira Eisaloo Gharghasheh

1 Introduction

Threat hunting is a field of cybersecurity which involves actively searching for threats in a system or network [1–5]. It often involves detecting anomalies among normal behavior patterns for a particular environment, which makes threat hunting a task which machine learning models are well-suited to [6–8]. When new, advanced machine learning algorithms are invented, cybersecurity researchers seek to apply them to improve automated security monitoring platforms [9–11]. With any security monitoring system, for the majority of the time there will be no threat present, and this makes it difficult to train supervised machine learning models to detect threats [12]. While a good system would of course be able to detect threats that have been seen before, for automated threat detection a model needs to be able to detect new threats [13, 14]. Hackers invent new ways to attack systems every day. Thus, a main challenge posed by automated security monitoring is the fact that threats are rare and therefore hard to define [15–18]. It is not sufficient to simply log the patterns created by previous attacks and keep an eye out for them in the future. Security monitoring systems need to be able to detect never-before-seen behavior that is a result of an attack [19]. While identifying abnormal behavior in a system is currently the best method for the automated detection of threats, it should be noted that this technique typically results in very high False Positive Rates (because not all abnormal behavior will indicate malicious activity) [20]. Due to the fact that it is impossible to train a threat hunting model to know exactly what a particular threat will look like, unsupervised learning [21] is preferable to supervised learning for automated security monitoring platforms [22]. There are exceptions to this, e.g.

K. Marsh (✉) · S. E. Gharghasheh
School of Computer Science, University of Guelph, Guelph, ON, Canada
e-mail: kmarsh08@uoguelph.ca; samira@cybersciencelab.org

© Springer Nature Switzerland AG 2021
K.-K. R. Choo, A. Dehghantanha (eds.), *Handbook of Big Data Analytics and Forensics*, https://doi.org/10.1007/978-3-030-74753-4_8

117

for malware detection it is beneficial for a machine learning model to learn what common malwares look like [23, 24]. In [25], fuzzy logic (unsupervised machine learning) is combined with Bayesian inference in order to create an optimized fuzzy model which is tested on a randomly generated dataset. Fuzzy logic involves making predictions based on predefined rules which are created based on prior knowledge about data distribution [26], while Bayesian inference tries to find the true probability of different predictions [27]. Fuzzy Bayesian learning combines advantages of both methods; the fuzzy model incorporates prior knowledge about the data in question, and an iterative Bayesian method known as Markov Chain Monte Carlo (MCMC) optimizes the parameters of the fuzzy model [28]. In this paper, we transformed the code that was provided by [25] and use it for cyber threat hunting [29–37]. We used three cybersecurity datasets, with one containing opcode samples and the other two containing monitoring data. Each dataset contains both normal and attack samples, and the success of the fuzzy Bayesian model will be based on its ability to detect attack samples. We redesigned the base fuzzy model for each of our three datasets. MCMC code then optimized the parameters of each fuzzy model, and finally the optimized parameters could be loaded back into each respective fuzzy model (creating the finalized fuzzy Bayesian model). For evaluation measures we used accuracy, True Positive Rate (TPR), False Positive Rate (FPR), F1 score, precision, and Area Under Curve (AUC).

This work seeks to take a novel machine learning algorithm, fuzzy Bayesian learning, and investigate its applicability to threat hunting. As mentioned earlier, cybersecurity researchers are always looking for new machine learning algorithms that are better adapted to the difficult task of system anomaly detection [9, 10, 38–42]. Potential contributions for future work include details on the requirements of a dataset to be suited for fuzzy Bayesian learning, as well as suggestions for developing the base fuzzy model (before MCMC).

Section 2 of this paper contains a literature review on similar papers from recent years. Section 3 details the methodology used in this work, Sect. 4 contains the results of our experiments, and in Sect. 5 we draw conclusions and suggest future works. Finally, references are in last section.

2   Related Works

A fuzzy probability Bayesian network (FBPN) is proposed in [43] for dynamic risk assessment. They used this system to detect and predict risks in Industrial Control Systems (ICSs). In this paper, for evaluating cybersecurity risk on ICSs they developed an approximate dynamic inference method and implemented a noise filter to achieve better results. In [44] they propose a combination of fuzzy hashing and augmented YARA rules to attain better results in ransomware triaging (assigning levels of threat to different types). The results showed that their approach is slightly better for ransomware triaging than just using standard YARA rules, with an accuracy of 98.21%. In [45], the researchers proposed an anomaly detection system by combining fuzzy c-means clustering and artificial neural networks in cloud environments. The performance of their suggested system was 92.73% for precision, 99.12% for recall and finally 96.31% for F-value. A two-step solution for malware detection on mobile devices by combining a Naive Bayesian model and the fuzzy c-means algorithm has been introduced in [46] By their new method, they achieved an accuracy of 99.9% and a faster average speed in malware detection compared to existing models. In order to detect ransomware and determine their families, [1] proposed a Deep Ransomware Threat Hunting and Intelligence System (DRTHIS). They achieved an F-measure of 99.6% and a TPR of 97.2% for ransomware classification. The authors in [47] have proposed a technique of combining the Bayesian and Dempster-Shafer theory (BDST) to compute trust for the delivery of packets at the node level. Additionally, they combined BDST with fuzzy theory to compute trust for secured routing at the link level and proposed "fuzzy- based Bayesian Dempster–Shafer trusted routing (BDSFTR)". Considering a network where 10% of nodes are faulty, the highest packet delivery ratio achieved for BDST and BDSFTR was 91.6%. In [23], they have implemented fast fuzzy pattern tree and fuzzy models to classify malware based on opcode patterns. They achieved an accuracy of 100% on an IoT dataset using potential heuristics on a fast fuzzy pattern tree. In [48] a novel anti-phishing method was proposed. They developed new features using hybrid feature analysis and obtained rules to create a fuzzy model. Using the fuzzy model, they attained an accuracy of 93% in phishing detection. A Two-hidden-layered Extreme Learning Machine has been proposed in [49] to detecting and analyzing malware. They achieved 99.65% accuracy in detecting malware on the IoT dataset. In [50], a comprehensive survey on the security of IoT, challenges and potential solutions was proposed. To avoid security threats in IoT, they classified IoT security challenges and their equivalent solutions by the layered architecture. A novel Non-Interactive Zero-Knowledge Proofs (NIZKP) has been proposed in [51] for authenticating the IoT devices.

While the traditional ZKP uses graph isomorphism their approach was based on Merkle trees. In [52], for detecting IoT malware and polymorphic malware they combined sequential pattern mining with machine learning algorithms. They were able to achieve more than 99% accuracy and F-measure. A deep Recurrent Neural Network was proposed in [9] to detecting malware on the IoT. They achieve the highest accuracy of 99.18% in malware detection with Long Short-Term Memory (LSTM) configuration [53]. has addressed the forensic and security challenges in the IoT environment. They also introduced the papers that proposed corresponding solutions. A comprehensive literature review on machine learning security solutions for the IoT and its structure has been introduced in [54]. Their survey contained all papers regards to IoT and machine learning security solutions of IoT up to 2019.

3   Methodology

This section will describe the process that was used to transform the code from [26] so that the fuzzy Bayesian model could be applied to cybersecurity datasets to detect threats. The Bayesian part of the model remained the same; its goal is to run many iterations of MCMC in order to optimize the parameters of the base fuzzy model. The base fuzzy model needed to be entirely redesigned to suit other datasets, due to the rule-based nature of fuzzy logic. A description is provided for each of the datasets used in this paper, followed by a description of the base code which was adapted for threat hunting. Finally, the process for redesigning the base fuzzy model and subsequent testing on cybersecurity datasets is provided.

3.1   Description of Cybersecurity Datasets

Three datasets were used to test the threat hunting abilities of the fuzzy Bayesian model designed in [26]. The first dataset consisted of opcode samples from IoT devices; some samples were benign (representing normal activity on the device) and some samples were malicious (representing malware activity on the device). The second dataset contains monitoring samples from a Secure Water Treatment (SWAT) system that was designed for research purposes, with samples being collected during both normal circumstances and attack circumstances (when the system is being attacked by a hacker) [55]. The third dataset has been generated through simulation of a C-Town water distribution system using a MATLAB toolbox [56] The dataset mostly contains data from normal activity in the water distribution system, but also includes samples from when the system was under attack.

3.2   Description of Base Code

The base code that was provided in [25] contains three main components for running the fuzzy Bayesian pipeline. The first component creates the base fuzzy model using rules that are provided in the code. After initiating the fuzzy model, the dataset of interest is loaded, and the output value of the fuzzy model is computed for each entry in the dataset. The input values and corresponding output values are then saved to a MATLAB-style file. This MATLAB file is then loaded into the second component of the pipeline, where 10,000 iterations of MCMC are performed to obtain optimized parameters for the base fuzzy model. Each iteration of MCMC creates a new prediction for optimized parameters, and these predictions are saved in an outputted CSV file. Each parameter will be represented by a column in the CSV file, and the mean can be taken from each column to obtain the optimized

Fig. 1 An overview of the base code. The third component is replaced in this paper

parameters (discarding predictions near the beginning and end of MCMC of the 10,000 iterations).

The final component provided in [25] loads the optimized parameters into the base fuzzy model and then plots the posterior distribution of the fuzzy membership functions. The plot is obtained by computing the fuzzy output for each individual parameter estimation that was produced by MCMC, for each membership function. For the purpose of this paper, we were not interested in plotting the posterior distribution of membership functions but instead sought to observe if the optimized parameters resulted in higher accuracy of the fuzzy model. We replaced this component with one that initializes the base fuzzy model using the mean of the MCMC columns as the parameters, and then runs that model on the dataset to obtain evaluation metrics.

An overview of the base code is illustrated in Fig. 1.

3.3 Designing Fuzzy Models

For the first component of the fuzzy Bayesian system, the base fuzzy model had to be recreated for each cybersecurity dataset. Because of this, the final code for this paper is divided into three different pipelines, one for each dataset. The difficulty of designing fuzzy models rests within defining meaningful rules. In many cases where fuzzy logic is employed to address real-world problems, the rules are defined and/or optimized based on the knowledge of experts in the field(s) relating to the dataset of interest [57]. For example, a fuzzy model for diagnosing disease in patients is likely to perform better when incorporating a doctor's opinion into the diagnosis. For this paper, we have three cybersecurity datasets and for each one we wish to look for anomalies. Due to the nature of anomaly detection, it would be difficult to incorporate an expert's opinion into our fuzzy model. Instead, we performed feature selection and reduction on the datasets and investigated the distribution of the final feature sets. The IoT dataset required additional preprocessing whereas the BATADAL and SWAT datasets were provided in a format that could be immediately fed into machine learning algorithms.

3.3.1 Feature Extraction: IoT Dataset

For the IoT malware dataset, initial data was provided in the form of many opcode samples which were labelled as benign or malicious (malicious samples representing malware). To convert these opcode samples into a usable dataset with features, frequently occurring patterns needed to be extracted. Each pattern is a feature. A pattern consists of a subset of operation codes, and the frequency of patterns in each sample become the feature values. Patterns which occur many times in a file provide more insight for classifying samples as benign or malicious. Therefore, the minimum frequency to be considered a pattern was set to 50 (pattern must appear at least 50 times in a file to be recognized as a pattern), while the length of a pattern was set to 3 (3 consecutive opcodes). This resulted in a total number of 4122 unique patterns, and therefore a new dataset was created where each sample/file has 4122 associated features.

3.3.2 Feature Reduction & Selection

For the BATADAL and SWAT datasets, the number of initial features were 45 and 79, respectively. For all 3 datasets, we reduced the total number of features to 2 in order to lower the overall complexity of the fuzzy model. This is because for each input feature, the fuzzy model needs to account for 3 possible instances (fuzzy values) for that feature: low, medium, and high. Therefore, with just two input features, 9 fuzzy rules are required to cover each possible combination of fuzzy values: low-low, low-medium, low-high, medium-low, medium-medium, medium-high, high-low, high-medium, and high-high. In order to reduce 45+ features to 2 features which can still be used to make predictions about the data, an embedding technique was used on each of the 3 datasets.

Due to the high-dimensionality of the IoT dataset that was produced in Sect. 3.3.1, univariate feature selection was used to reduce the 4122 features to a more manageable 200. The 200 features with the highest ANOVA F-values were kept. Following this, t-distributed Stochastic Neighbor Embedding (t-SNE) was used on all datasets to transform each to have only 2 features. t-SNE is an algorithm that helps with visualization of high-dimensional datasets using feature reduction [58], which is useful for the purpose of creating rules for fuzzy logic. Several different embedding techniques were tried on each dataset individually, and t-SNE produced the highest amount of separability between benign and malicious data in each case.

3.3.3 Designing Fuzzy Rules Based on Distribution of Features

Once the datasets were transformed to 2 features, creating fuzzy rules to classify samples as benign or malicious was a matter of percentages. Fuzzy models are created with antecedents and consequents. The antecedents take input values in crisp form and fuzzify them, then the model applies fuzzy rules. Finally, the consequents

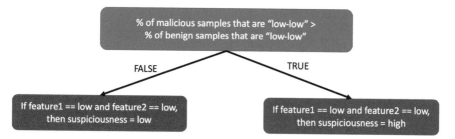

Fig. 2 Illustration of how the fuzzy rule for the low-low category would be determined. This process would be repeated for the remaining 8 fuzzy categories (low-medium, low-high, etc.)

convert fuzzy values back into crisp values for output. Therefore, we had two antecedents (one for each input feature) and one consequent which was labelled "suspiciousness", i.e. a high output value indicates the sample in question is highly suspected to be malicious. For each of our 2 antecedents, 3 membership functions were assigned for fuzzy values: low, medium, and high. 9 fuzzy rules were created to account for possible combinations of the two features (see Sect. 3.3.2). To simplify membership functions, data was normalized so that each feature was between 0 and 1. An example of a low-low sample would be one where both input features are below 0.25.

To determine the fuzzy rules, the percentage of benign samples in each combination category was compared to the percentage of malicious samples in each matching category. For example, if 60% of malicious samples fell into the low-low category (both features fuzzified to "low") but only 10% of benign samples fell into the same category, the corresponding rule would be: "If feature1 is low and feature2 is low, then suspiciousness is high". This is illustrated in Fig. 2.

3.4 MCMC and Testing

Once the base fuzzy model was created, the base code from [25] takes over running 10,000 iterations of MCMC to obtain optimized parameters. After MCMC, we re-created our base fuzzy model but with the parameters of the membership functions being replaced by those outputted from MCMC. Finally, we simply run the optimized fuzzy model on the correlating dataset and obtain evaluation measurements. To evaluate the performance of the fuzzy models we used accuracy, TPR, FPR, F1 scores, precision, and AUC.

4   Results & Discussion

This section highlights the results that we achieved after implementing an FBL algorithm to classify 3 different cybersecurity datasets. The evaluation metrics which were specified in Sect. 3.4 are used to evaluate performance both before and after each fuzzy model has been optimized (although pre-optimization metrics will only include accuracy, TPR, and FPR). Results for each dataset will be discussed in 3 different subsections, and a summary of results can be seen in Table 1. Finally, a brief evaluation is given for the overall performance of FBL on our datasets.

4.1   IoT Dataset

The best results we received were for the IoT dataset, which is no doubt due to the high separability of the data between classes. The high separability is in turn a result of the fact that malware samples, even if representing different malwares which were created by different people, tend to have a lot in common in terms of frequent opcode patterns [52]. Therefore, when executing the technique described in Sect. 3.3.3 for calculating percentages of benign/malicious samples in each fuzzy category, there were strong connections between each category and a particular class. For example, 61% of benign samples fell into the low-low category whereas 0% of malicious samples were in that category. As a result, the following rule could be created with exceptional confidence: "If feature1 is low and feature2 is low, suspiciousness is low". Confidence was similarly high for all rules that were created for the IoT dataset.

The initial performance of the fuzzy model created for the IoT dataset was excellent, with an accuracy of 97.97%, TPR of 98.52%, and FPR of 2.58%. After optimization of the parameters for the fuzzy membership functions, the accuracy rose slightly to 98.15%. While the FPR decreased to 1.49%, unfortunately the TPR also decreased to 97.80%. While a low FPR is ideal, with malware classification we are much more concerned with achieving a TPR that is close to 100% so that the amount of malware that goes undetected is minimal. Therefore, for the purpose

Table 1 Summary of results for final FBL model

	IoT dataset	SWAT dataset	BATADAL dataset
Accuracy	98.15% ↑0.18%	76.43% ↑3.77%	67.39% ↓0.05%
TPR	97.80% ↓0.72%	74.43% ↓6.13%	4.47% ↓25.81%
FPR	1.49% ↓1.09%	21.21% ↓14.03%	3.49% ↓27.6%
F1 score	98.16%	77.34%	7.98%
Precision	98.52%	80.52%	37.19%
AUC	98.09%	84.63%	53.08%

Accuracy, TPR, and FPR results show the increase/decrease from the base fuzzy model of malware classification, the base fuzzy model performs better before the Bayesian technique of MCMC is involved. However, it should be noted that MCMC did in fact succeed at slightly increasing accuracy by reducing FPR, and there may be certain research circumstances where a lower FPR is more important than a high TPR. F1 score, precision, and AUC metrics for the optimized model can be found in Table 1.

4.2 SWAT Dataset

Unlike the IoT dataset, where malware samples had distinct behavioral patterns which could be used to easily identify them, the SWAT dataset was not as easy in terms of drawing a "line" between benign and malicious samples. This is because the malicious samples in the SWAT dataset represent activity monitoring samples from a system when the system is under attack by a hacker. Therefore, not only are the malicious samples exceedingly outnumbered by the benign samples, they are only uniquely identifiable in the sense that they appear slightly out of the norm compared to benign activity samples. The same technique for creating fuzzy rules was used as for the IoT dataset, however there was much less confidence in the rules. For example, 28% of benign samples may fall into the low-low category, but 20% of malicious samples fall into that category as well. The corresponding fuzzy rule would still be that a low-low sample has low suspiciousness, but now this will result in some incorrect classifications.

The initial accuracy of the fuzzy model for the SWAT dataset was 72.66%, with a TPR of 80.56% and an FPR of 35.24%. After optimization of parameters by MCMC, accuracy increased to 76.43% and FPR was reduced to 21.21%, but TPR was reduced to 74.43%. Unfortunately, the final results are similar to those of IoT in the sense that while accuracy and FPR are improved, the TPR is reduced and this is highly undesirable for a threat hunting algorithm. When threat hunting through anomaly detection, false positives are expected [59] whereas false negatives need to be avoided as much as possible. Once again, the FBL algorithm succeeded in the sense that it improved the accuracy of the base fuzzy model, but at the cost of a slightly lowered TPR which is an unacceptable consequence for threat hunting purposes. F1 score, precision, and AUC metrics for the optimized model can be found in Table 1.

4.3 BATADAL Dataset

Results for the BATADAL dataset were overall poor. Unfortunately, this dataset just did not conform well to any of the techniques that were described in the methodology. t-SNE still proved to be the best embedding technique to transform the data into a 2-feature dataset, however there was heavy overlap in the percentages of benign and malicious samples which fell into each fuzzy category. For example, for the low-low category, there may be roughly the same percentage of benign samples in the category as there are malicious. This made it extremely difficult to define fuzzy rules and led to several rules where the outputted fuzzy value for suspiciousness was medium instead of high or low. As a result, the base fuzzy model performed poorly.

The accuracy of the base fuzzy model was 67.44%, with a TPR of 30.28% and an FPR of 31.09%. Clearly, these results are unacceptable for the purpose of threat hunting, in particular the TPR. The 10,000 iterations of MCMC did not help the model; using the "optimized" parameters that were outputted, accuracy was actually **reduced** to 67.39%, with a TPR of 4.47% and an FPR of 3.49%. This suggests that this dataset is simply not suited for the FBL algorithm, which is likely consequent to the fact that this dataset is not suited for fuzzy logic.

One possible reason for the difficulty of creating a base fuzzy model for the BATADAL dataset could be that the number of malicious samples is far too outnumbered by the number of benign samples. There are 12,446 malicious samples in the BATADAL dataset compared to only 492 benign. It is possible that the collection of additional data which includes malicious samples would allow for a more accurate fuzzy model.

4.4 Evaluation of FBL

For the purposes of threat hunting, FBL proved very effective for the IoT dataset and moderately effective for the SWAT dataset. An accuracy of approximately 98% was achieved for the IoT dataset, which is the same accuracy reported in [9] where a deep Recurrent Neural Network was used to classify malware in a similar dataset. In [60], a deep learning model is used on the same SWAT dataset and a TPR of approximately 85% is achieved, which is higher than our 80.56% TPR (achieved with the base fuzzy model, pre-optimization). However, the deep learning model that is used in [60] is a Convolutional Neural Network, an algorithm which is significantly more computationally expensive than a fuzzy model.

For both IoT and SWAT datasets, the fuzzy model on its own performed well and in fact the Bayesian part of FBL (MCMC iterations) seemed to hurt the performance of classifying benign and malicious samples. While accuracy rose slightly in both cases, TPR was lowered and this is unacceptable for threat hunting purposes. However, for both cases, MCMC increased accuracy in part by reducing FPR. There may be scenarios where a low FPR is more important than a high TPR, in which case FBL would be a good way to optimize a fuzzy model.

Due to poor results for the BATADAL dataset, even for the base fuzzy model, these results were not used for evaluation of FBL.

5 Conclusion & Future Work

FBL proved to be an effective algorithm for threat hunting in 2 out of 3 cybersecurity datasets used in this work. For the BATADAL dataset, a base fuzzy model could not be built which was capable of decent performance; this could possibly be remedied by additional data collection which includes malicious samples. For the IoT and SWAT datasets, the algorithm successfully took a base fuzzy model for detecting malicious samples and optimized the parameters through 10,000 iterations of MCMC, which slightly improved the accuracy of each model. The accuracies of the initial fuzzy models for the IoT and SWAT datasets were 97.97% and 72.66%, respectively. The accuracies of the optimized fuzzy models for the IoT and SWAT datasets were 98.15% and 76.43%, respectively. Unfortunately, for both the IoT and the SWAT dataset, the increase in accuracy came at the cost of a slightly lowered TPR. This indicates that the algorithm is not entirely suited to threat hunting, as it is always desirable to detect as many real threats as possible. Even the best threat hunting algorithms are bound to have a high number of false positives, which means the lower FPR produced by parameter optimization in this paper is not beneficent.

While the parameter optimization segment of the FBL algorithm did not seem to assist with threat hunting, the base fuzzy models created in this paper showed promising results for the IoT and SWAT datasets. For future work, continued investigation should be performed on the use of fuzzy logic for threat hunting. Furthermore, while MCMC had little effect on fuzzy model performance in this paper, other methods which could potentially improve the TPR of threat hunting fuzzy models should be explored.

References

1. S. Homayoun et al., Deep dive into ransomware threat hunting and intelligence at fog layer. Futur. Gener. Comput. Syst. **90**(Jan 19), 94–104 (2018)
2. A.N. Jahromi, S. Hashemi, A. Dehghantanha, R.M. Parizi, K.-K.R. Choo, An enhanced stacked LSTM method with no random initialization for malware threat hunting in safety and time-critical systems. IEEE Trans. Emerg. Top. Comput. Intell. **4**(5), 630–640 (2020). https://doi.org/10.1109/tetci.2019.2910243
3. M.M. BehradFar et al., RAT hunter: Building robust models for detecting remote access trojans based on optimum hybrid features, in *Handbook of Big Data Privacy*, (Springer, Cham, 2020), pp. 371–383. https://doi.org/10.1007/978-3-030-38557-6_18
4. H. Darabian et al., A multiview learning method for malware threat hunting: Windows, IoT and android as case studies. World Wide Web **23**(2), 1241–1260 (2020)
5. A. Yazdinejad, R.M. Parizi, A. Dehghantanha, K.-K.R. Choo, P4-to-blockchain: A secure blockchain-enabled packet parser for software defined networking. Comput. Secur. **88** (2020). https://doi.org/10.1016/j.cose.2019.101629
6. S. Schmitt, F.I. Kandah, D. Brownell, Intelligent threat hunting in software-defined networking, in *2019 IEEE International Conference on Consumer Electronics (ICCE)*, (2019), pp. 1–5
7. H.H. Pajouh, R. Javidan, R. Khayami, D. Ali, K.-K.R. Choo, A two-layer dimension reduction and two-tier classification model for anomaly-based intrusion detection in IoT backbone networks. IEEE Trans. Emerg. Top. Comput. (2016). https://doi.org/10.1109/TETC.2016.2633228

8. A. Azmoodeh, A. Dehghantanha, M. Conti, K.-K.R. Choo, Detecting crypto-ransomware in IoT networks based on energy consumption footprint. J. Ambient. Intell. Humaniz. Comput. **9**(4), 1141–1152 (2018)
9. H. HaddadPajouh, A. Dehghantanha, R. Khayami, K.-K.R. Choo, A deep recurrent neural network based approach for internet of things malware threat hunting. Futur. Gener. Comput. Syst. **85**, 88–96 (2018). https://doi.org/10.1016/j.future.2018.03.007
10. D. Karev, C. McCubbin, R. Vaulin, Cyber threat hunting through the use of an isolation forest, in *Proceedings of the 18th International Conference on Computer Systems and Technologies*, (2017), pp. 163–170
11. A. Yazdinejad, R.M. Parizi, A. Dehghantanha, K.-K.R. Choo, Blockchain-enabled authentication handover with efficient privacy protection in SDN-based 5G networks, in *IEEE Transactions on Network Science and Engineering*, (IEEE, 2019), pp. 1–1. https://doi.org/10.1109/tnse.2019.2937481
12. H. Karimipour, A. Dehghantanha, R.M. Parizi, K.-K.R. Choo, H. Leung, A deep and scalable unsupervised machine learning system for cyber-attack detection in large-scale smart grids. IEEE Access **7**, 80778–80788 (2019)
13. M.N.S. Miazi, M.M.A. Pritom, M. Shehab, B. Chu, J. Wei, The design of cyber threat hunting games: A case study, in *2017 26th International Conference on Computer Communication and Networks (ICCCN)*, (2017), pp. 1–6
14. H. Hashemi, A. Azmoodeh, A. Hamzeh, S. Hashemi, Graph embedding as a new approach for unknown malware detection. J. Comput. Virol. Hacking Tech. **13**(3), 153–166 (2017)
15. J. Sakhnini, H. Karimipour, A. Dehghantanha, R.M. Parizi, G. Srivastava, Security aspects of Internet of Things aided smart grids: A bibliometric survey. Internet of things **1**, 100111 (2019)
16. P.N. Bahrami, A. Dehghantanha, T. Dargahi, R.M. Parizi, K.-K.R. Choo, H.H.S. Javadi, Cyber kill chain-based taxonomy of advanced persistent threat actors: Analogy of tactics, techniques, and procedures. J. Inf. Process. Syst. **15**(4), 865–889 (2019)
17. S. Grooby, T. Dargahi, A. Dehghantanha, Protecting IoT and ICS platforms against advanced persistent threat actors: Analysis of APT1, Silent Chollima and molerats, in *Handbook of Big Data and IoT Security*, (Springer, Berlin, 2019), pp. 225–255
18. H. Mwiki, T. Dargahi, A. Dehghantanha, K.-K.R. Choo, Analysis and triage of advanced hacking groups targeting western countries critical national infrastructure: APT28, RED October, and Regin, in *Critical Infrastructure Security and Resilience*, (Springer, Berlin, 2019), pp. 221–244
19. A. Al-Abassi, H. Karimipour, A. Dehghantanha, R.M. Parizi, An ensemble deep learning-based cyber-attack detection in industrial control system. IEEE Access **8**, 83965–83973 (2020)
20. A. Sharma, Z. Kalbarczyk, J. Barlow, R. Iyer, Analysis of security data from a large computing organization, in *2011 IEEE/IFIP 41st International Conference on Dependable Systems & Networks (DSN)*, (2011), pp. 506–517
21. A. Azmoodeh, A. Dehghantanha, R.M. Parizi, S. Hashemi, B. Gharabaghi, G. Srivastava, Active spectral botnet detection based on eigenvalue weighting, in *Handbook of Big Data Privacy*, (Springer, Cham, 2020), pp. 385–397. https://doi.org/10.1007/978-3-030-38557-6_19
22. H. Karimipour, H. Leung, Relaxation-based anomaly detection in cyber-physical systems using ensemble Kalman filter. IET Cyber-Physical Syst. Theory Appl. **5**(1), 49–58 (2020)
23. E.M. Dovom, A. Azmoodeh, A. Dehghantanha, D.E. Newton, R.M. Parizi, H. Karimipour, Fuzzy pattern tree for edge malware detection and categorization in IoT. J. Syst. Archit. **97**, 1–7 (2019)
24. M. Alaeiyan, A. Dehghantanha, T. Dargahi, M. Conti, S. Parsa, A multilabel fuzzy relevance clustering system for malware attack attribution in the edge layer of cyber-physical networks. ACM Trans. Cyber-Physical Syst. **4**(3), 1–22 (2020)
25. I. Pan, D. Bester, Fuzzy Bayesian learning. IEEE Trans. Fuzzy Syst. **26**(3), 1719–1731 (2017)
26. L.A. Zadeh, Fuzzy logic. Computer (Long Beach Calif) **21**, 83–93 (1988)
27. A.P. Dempster, A generalization of Bayesian inference. J. R. Stat. Soc. Ser. B **30**(2), 205–232 (1968)

28. C. Andrieu, N. De Freitas, A. Doucet, M.I. Jordan, An introduction to MCMC for machine learning. Mach. Learn. **50**(1–2), 5–43 (2003)
29. A. Yazdinejad, G. Srivastava, R.M. Parizi, A. Dehghantanha, H. Karimipour, S.R. Karizno, SLPoW: Secure and low latency proof of work protocol for blockchain in green IoT networks, in *2020 IEEE 91st Vehicular Technology Conference (VTC2020-Spring)*, (2020), pp. 1–5
30. A. Yazdinejad, R.M. Parizi, A. Dehghantanha, K.-K.R. Choo, Blockchain-enabled authentication handover with efficient privacy protection in SDN-based 5G networks. IEEE Trans. Netw. Sci. Eng. (2019). https://doi.org/10.1109/TNSE.2019.2937481
31. A. Singh, K. Click, R.M. Parizi, Q. Zhang, A. Dehghantanha, K.-K.R. Choo, Sidechain technologies in blockchain networks: An examination and state-of-the-art review. J. Netw. Comput. Appl. **149**, 102471 (2020). https://doi.org/10.1016/j.jnca.2019.102471
32. A. Yazdinejad, R.M. Parizi, A. Dehghantanha, Q. Zhang, K.-K.R. Choo, An energy-efficient SDN controller architecture for IoT networks with blockchain-based security. IEEE Trans. Serv. Comput. (2020). https://doi.org/10.1109/TSC.2020.2966970
33. D. Połap, G. Srivastava, A. Jolfaei, R.M. Parizi, Blockchain technology and neural networks for the internet of medical things, in *IEEE INFOCOM 2020 – IEEE Conference on Computer Communications Workshops (INFOCOM WKSHPS)*, (2020), pp. 508–513. https://doi.org/10.1109/INFOCOMWKSHPS50562.2020.9162735
34. A. Yazdinejad, G. Srivastava, R.M. Parizi, A. Dehghantanha, K.-K.R. Choo, M. Aledhari, Decentralized authentication of distributed patients in hospital networks using blockchain. IEEE J. Biomed. Health Inform. **24**(8), 2146–2156 (2020)
35. Q. Chen, G. Srivastava, R.M. Parizi, M. Aloqaily, I. Al Ridhawi, An incentive-aware blockchain-based solution for internet of fake media things. Inf. Process. Manag., 102370 (2020). https://doi.org/10.1016/j.ipm.2020.102370
36. A. Yazdinejad, R.M. Parizi, A. Bohlooli, A. Dehghantanha, K.-K.R. Choo, A high-performance framework for a network programmable packet processor using P4 and FPGA. J. Netw. Comput. Appl. **156**, 102564 (2020)
37. R.M. Parizi, S. Homayoun, A. Yazdinejad, A. Dehghantanha, K.-K.R. Choo, Integrating privacy enhancing techniques into blockchains using sidechains, in *IEEE Canadian Conference of Electrical and Computer Engineering, CCECE 2019*, (2019). https://doi.org/10.1109/CCECE.2019.8861821
38. A. Yazdinejad, R.M. Parizi, G. Srivastava, A. Dehghantanha, K.-K.R. Choo, Energy efficient decentralized authentication in internet of underwater things using blockchain, in *2019 IEEE Globecom Workshops (GC Wkshps)*, (2019), pp. 1–6
39. V. Mothukuri, R.M. Parizi, S. Pouriyeh, Y. Huang, A. Dehghantanha, G. Srivastava, A survey on security and privacy of federated learning. Futur. Gener. Comput. Syst. **115**, 619–640 (2020)
40. A. Yazdinejad, H. HaddadPajouh, A. Dehghantanha, R.M. Parizi, G. Srivastava, M.-Y. Chen, Cryptocurrency malware hunting: A deep recurrent neural network approach. Appl. Soft Comput. Elsevier **96**, 106630 (2020)
41. M. Aledhari, R. Razzak, R.M. Parizi, F. Saeed, Federated learning: A survey on enabling technologies, protocols, and applications. IEEE Access **8**, 140699–140725 (2020). https://doi.org/10.1109/ACCESS.2020.3013541
42. A. Yazdinejad, R.M. Parizi, A. Dehghantanha, H. Karimipour, G. Srivastava, M. Aledhari, Enabling drones in the internet of things with decentralized blockchain-based security. IEEE Internet Things J., 1 (2020). https://doi.org/10.1109/jiot.2020.3015382
43. Q. Zhang, C. Zhou, Y.-C. Tian, N. Xiong, Y. Qin, B. Hu, A fuzzy probability Bayesian network approach for dynamic cybersecurity risk assessment in industrial control systems. IEEE Trans. Ind. Inform. **14**(6), 2497–2506 (2017)
44. N. Naik, P. Jenkins, N. Savage, L. Yang, K. Naik, J. Song, Augmented YARA rules fused with fuzzy hashing in ransomware triaging, in *2019 IEEE Symposium Series on Computational Intelligence (SSCI)*, (2019), pp. 625–632
45. N. Pandeeswari, G. Kumar, Anomaly detection system in cloud environment using fuzzy clustering based ANN. Mob. Netw. Appl. **21**(3), 494–505 (2016)

46. A. Razaque, Z. Xihao, W. Liangjie, M. Almiani, Y. Jararweh, M.J. Khan, Naïve Bayesian and fuzzy C-means algorithm for mobile malware detection precision, in *2018 Fifth International Conference on Internet of Things: Systems, Management and Security*, (2018), pp. 239–243

47. D. Velusamy, G.K. Pugalendhi, Fuzzy integrated Bayesian Dempster–Shafer theory to defend cross-layer heterogeneity attacks in communication network of Smart Grid. Inf. Sci. (NY) **479**, 542–566 (2019)

48. R. AlShboul, F. Thabtah, N. Abdelhamid, M. Al-Diabat, A visualization cybersecurity method based on features' dissimilarity. Comput. Secur. **77**, 289–303 (2018)

49. A.N. Jahromi et al., An improved two-hidden-layer extreme learning machine for malware hunting. Comput. Secur. **89**, 101655 (2020)

50. H. HaddadPajouh, A. Dehghantanha, R.M. Parizi, M. Aledhari, H. Karimipour, A survey on internet of things security: Requirements, challenges, and solutions. Internet of Things **3**, 100129 (2019)

51. M. Walshe, G. Epiphaniou, H. Al-Khateeb, M. Hammoudeh, V. Katos, A. Dehghantanha, Non-interactive zero knowledge proofs for the authentication of IoT devices in reduced connectivity environments. Ad Hoc Netw. **95**, 101988 (2019)

52. H. Darabian, A. Dehghantanha, S. Hashemi, S. Homayoun, K.R. Choo, An opcode-based technique for polymorphic Internet of Things malware detection. Concurr. Comput. Pract. Exp. **32**(6), e5173 (2020)

53. M. Conti, A. Dehghantanha, K. Franke, S. Watson, Internet of Things security and forensics: Challenges and opportunities. Futur. Gener. Comput. Syst. **78**, 544–546 (2018). https://doi.org/10.1016/j.future.2017.07.060

54. S.M. Tahsien, H. Karimipour, P. Spachos, Machine learning based solutions for security of Internet of Things (IoT): A survey. J. Netw. Comput. Appl. **161**, 102630 (2020)

55. J. Goh, S. Adepu, K.N. Junejo, A. Mathur, A dataset to support research in the design of secure water treatment systems, in *International Conference on Critical Information Infrastructures Security*, (2016), pp. 88–99

56. R. Taormina et al., Battle of the attack detection algorithms: Disclosing cyber attacks on water distribution networks. J. Water Resour. Plan. Manag. **144**(8), 4018048 (2018)

57. A. Kaufmann, Theory of expertons and fuzzy logic. Fuzzy Sets Syst. **28**(3), 295–304 (1988)

58. L. van der Maaten, G. Hinton, Visualizing data using t-SNE. J. Mach. Learn. Res. **9**(Nov), 2579–2605 (2008)

59. L. Franklin, M. Pirrung, L. Blaha, M. Dowling, M. Feng, Toward a visualization-supported workflow for cyber alert management using threat models and human-centered design, in *2017 IEEE Symposium on Visualization for Cyber Security (VizSec)*, (2017), pp. 1–8

60. M. Kravchik, A. Shabtai, Detecting cyber attacks in industrial control systems using convolutional neural networks, in *Proceedings of the 2018 Workshop on Cyber-Physical Systems Security and Privacy*, (2018), pp. 72–83

Cyber-Attack Detection in Cyber-Physical Systems Using Supervised Machine Learning

Prabhat Semwal and Akansha Handa

1 Introduction

The cyber-physical systems can be defined as the systems built by integrating sensors, computers, networks, communication, and other digital monitoring components into physicals infrastructure to control or monitor the infrastructure remotely and autonomously [1–3]. Some real-world examples of CPS include Smart grids, medical monitoring systems, robotics, autonomous vehicles, soil treatment plants, and water treatment plants [4–8]. The cyber-physical infrastructure operations include both cyber and physical aspects which make these systems vulnerable to both cyber and physical security threats. The attack on CPS can have a huge impact due to the diversity and scope of operations of these structures [6, 9–13]. Thus, the cyber aspect of such CPS has been studied in many pieces of research, which contributed their finds in detecting the cyber-attacks on CPS using machine learning [12, 14–17]. Advancement in Machine Learning and Deep Learning models has motivated the cybersecurity communities for leveraging these models so as to enhance the privacy and security of CPS [18–24]. During the past decade several models have been proposed for a diverse range of cybersecurity including malware detection [25–28], threat hunting [29–32] and privacy protection [33].

In this paper, we have used the SWat dataset which is the data collected from a Secure Water Treatment plant [34]. The data was collected for both normal operational days and few days with attacks on the water treatment. The dataset is processed and used to perform the cyber-attack detection on CPS systems using different supervised machine learning algorithms. We have performed the comparative analysis on the four models based on the major evaluation matrices:

P. Semwal (✉) · A. Handa
School of Computer Science, University of Guelph, Guelph, ON, Canada
e-mail: psemwal@uoguelph.ca; ahanda@uoguelph.ca

© Springer Nature Switzerland AG 2021
K.-K. R. Choo, A. Dehghantanha (eds.), *Handbook of Big Data Analytics and Forensics*, https://doi.org/10.1007/978-3-030-74753-4_9

Accuracy, True Positive Rate (TPR), False Positive Rate (FPR) and the Receiver Operating Characteristics (ROC) curve and Area Under ROC Curve (AUC).

2 Literature Review

In earlier studies, many computer scientists have proposed various approaches to resolve cyber threat hunting problems using different techniques of machine learning [35–39]. Cyber-attack detection is usually accomplished by grouping using power device data or measurements [40–44]. The involvement of risks or attacks is measured in various security and contact levels of the network. Cyber-attacks are observed by measurements by the improved state- estimation techniques using mode-based technique [45]. Numerous studies have presented network traffic-based intrusion detection Ghaeini et al. [46] employ this approach on the SWaT dataset used in our study. Similarly, [47] proposed an Enhanced SVM approach with combined features from two machine learning techniques demonstrated a low false-positive rate. Another paper [48] uses the Random forest Algorithm and achieves a significant accuracy of 94.0187% for cyber-attack detection. A behavior-based machine learning (ML) approach for the detection of any abnormal behavior or attack that may attempt to modify the behavior of the CPS [15]. This method not only recognizes the cyber-attack occurred on a layer of the physical process, but it also identifies the specific attack type. In This study [49] learns how to combine different machine learning methods with the IDS improving the accuracy of threat identification. A prototype IDS is expected in this study. This IDS prototype is equipped to improve accuracy in the identification of several attacks through a combination of machine learning methods. This method not only recognizes the cyber-attack occurred on a layer of the physical process, but it also identifies the specific attack type. In [50] the proposed cyberattack detection system has high detection accuracy and wide attack coverage in order to detect unrecognized attacks using network and host system information.

3 Methodology

This section will describe the process followed to build our supervised machine learning models which can detect the cyber-attack samples from the SWat dataset.

3.1 Dataset Processing

The Swat dataset consists of 77 features and a total of 14,995 data points, 9521 normal and 5474 attack data points. The few features like timestamp and other less critical features were removed to process the dataset. The label feature (Target) was marked as 1 for attack and 0 for the normal activity data point.

3.1.1 Feature Selection

To reduce the overall dimensionality of the dataset, we performed the feature extraction process. The best feature that can contribute to the target variable was extracted by combining results of ExtraTreeClassifier and SelectKBest algorithms of Scikit-learn library and are shown in Fig. 1a, b respectively.

The most common and highest-ranked features were extracted and used for all the four-classification model. As shown in Table 1, the major operations of the water treatment plant was used as a major feature category set and the same category of features were used to identify the functionality of water plants at different levels process.

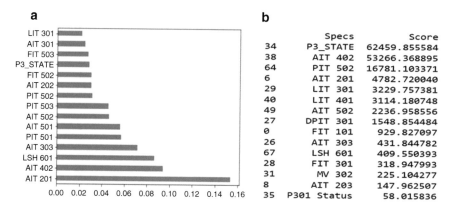

Fig. 1 (**a**) ExtraTreeClassifier (**b**) SelectKBest Result

Table 1 Feature category

Feature name	Description
FIT	Sensor: Inflow into the water tank
LIT	Sensor: Level transmitter
PIT	Sensor: Pressure meter
LSH	Alarm
AIT	Sensor

3.2 *Machine Learning Classifiers*

For the detection of cyber-attack samples, the KNN, SVM, Decision Tree, and Random Forest, classifiers were trained and tested on the transformed dataset, and results were recorded for comparative analysis.

3.2.1 KNN Model

In the KNN model, we used the processed dataset explained in Sect. 3.2.1. The KNN was implemented with the use of the Sckilearn library and in KNN we initialize the K = 4, but after trial and error K was finally set to 1 and the model was trained with K = 1 on the processed dataset.

3.2.2 SVM Model

The SVM model was trained and tested on the processed dataset. For SVM, kernel function was set to linear, and probability was set to True.

3.2.3 DT Model

Our DT Model was trained with the processed dataset. The DT model simply designs an inverted tree structure on the base of a trained dataset and then classify a sample by tracing the down designed tree.

3.2.4 RF Model

The RF model is like the DT model, but the RF model creates multiple decision trees instead of only one decision tree. In our RF model, the maximum depth was set to 2.

4 Results and Discussion

This section highlights the results achieved with different supervised machine learning techniques in detecting cyber-attack on a CPS system and will describe the comparative analysis results.

4.1 Evaluation Measures

To evaluate and compare the performance of the models, we have used the commonly used evaluation metrics. Table 2 contains a description of the used evaluation metrics for comparative analysis.

4.2 Experiment and Results

The processed dataset with the total samples of 14,994 and selected features was used to test all the models. All the models were trained on the processed dataset and the results observed on the basis of evaluation metrics (Table 2) are shown in Table 3.

4.3 Comparison of Models

In our experiment, the KNN model achieved an accuracy of 99%. The TPR received for KNN was 99.9% and the FPR was approximately 0%. With the SVM model, we received an accuracy of 98.7% and the average values of TPR and FPR were 99% and 0.01% respectively. Our DT model received 99% accuracy on the processed dataset and approximately 99% TPR and 0% FPR. Whereas, the RF model hit the accuracy of 96% with 98% APR and 0.01% FPR. According to the three-evaluation metrics values mentioned in Table 3, the DT model performed more effectively than other supervised machine learning models in classifying the cyber-attack samples in the Swat dataset.

Table 2 Description of evaluation metric used for comparative analysis

Evaluation metric	Description
TPR	$\frac{TP}{TP+FN}$
FPR	$\frac{FP}{FP+TN}$
Accuracy	$\frac{TP+TN}{TP+TN+FP+FN}$
ROC Curve	Formed by plotting TPR against FPR at various threshold settings
AUC	The area under the ROC curve

Table 3 Observed accuracy, TPR and FPR values

Model	Accuracy	TPR	FPR
KNN	99.65	99.9	0.008
SVM	98.70	99.03	0.018
DT	99.9	99.0	0.006
RF	96.3	98.6	0.013

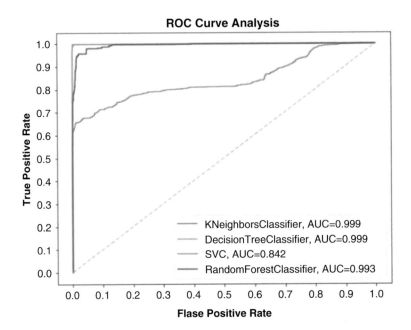

Fig. 2 ROC Curve for all four supervised machine learning models

4.4 ROC Curve

A ROC curve is a common graphical evaluation metric that is used for evaluating the performance of different machine learning classifiers. It allows us to analyze the binary classifier's capability of distinguishing between classes [51]. It is simply a plot of TPR and FPR at different threshold settings.

The ROC curve for classification of cyber-attack samples on a processed dataset, for all four supervised classification models, is shown in Fig. 2. As shown in the legend, the overall AUC value for KNN was 0.99, 0.99 for DT, 0.84 for SVM, and 0.99 for RF. The average AUC of all four models was extremely close to 1 which depicts that all four models perform better for binary classification of cyber-attack in a processed dataset. Although, the AUC value observed for all four models was close to 1. However, the AUC value for both KNN and DT is almost equal to 1 with AUC equal to 0.999 for both KNN and DT.

5 Conclusion

We were able to successfully design the four different machine learning models to classify the cyber-attack samples accurately from the Swat dataset. The results achieved using the critical evaluation metrics allowed us to perform effective comparative analysis and propose the most suitable algorithm. Using these four supervised machine learning algorithms, we achieved an overall accuracy of 99% with KNN, 98% with SVM, 99% with DT, and 96% with RF. On the base of all the evaluation metrics, the DT outperforms the other classifier models with a reasonable high accuracy of 99.9% and other almost ideal evaluation metrics.

The future work will be to evaluate the other supervised machine learning algorithms and to experiment with the different cyber-physical system datasets.

References

1. H. Karimipour, H. Leung, Relaxation-based anomaly detection in cyber-physical systems using ensemble Kalman filter. IET Cyber-Phys. Syst. Theory Appl. **5**(1), 49–58 (2020)
2. A. Yazdinejad, A. Bohlooli, K. Jamshidi, Efficient design and hardware implementation of the OpenFlow v1.3 Switch on the Virtex-6 FPGA ML605. J. Supercomput. **74**(3), 1299–1320 (2018). https://doi.org/10.1007/s11227-017-2175-7
3. A. Yazdinejad, R.M. Parizi, A. Dehghantanha, K.-K.R. Choo, P4-to-blockchain: A secure blockchain-enabled packet parser for software defined networking. Comput. Secur. **88**, 101629 (2020). https://doi.org/10.1016/j.cose.2019.101629
4. H. Karimipour, A. Dehghantanha, R.M. Parizi, K.-K.R. Choo, H. Leung, A deep and scalable unsupervised machine learning system for cyber-attack detection in large-scale smart grids. IEEE Access **7**, 80778–80788 (2019)
5. H. Karimipour, V. Dinavahi, Extended Kalman filter-based parallel dynamic state estimation. IEEE Trans. Smart Grid **6**(3), 1539–1549 (2015)
6. F. Daryabar, A. Dehghantanha, N.I. Udzir, S.B. Shamsuddin, Towards secure model for SCADA systems, in *Proceedings Title: 2012 International Conference on Cyber Security, Cyber Warfare and Digital Forensic (CyberSec)*, (2012), pp. 60–64
7. S. Walker-Roberts, M. Hammoudeh, A. Dehghantanha, A systematic review of the availability and efficacy of countermeasures to internal threats in healthcare critical infrastructure. IEEE Access **6**, 25167–25177 (2018, March). https://doi.org/10.1109/ACCESS.2018.2817560
8. S. Nakhodchi, A. Dehghantanha, H. Karimipour, Privacy and security in smart and precision farming: A bibliometric analysis, in *Handbook of Big Data Privacy*, (Springer, Cham, 2020), pp. 305–318
9. S. Watson, A. Dehghantanha, Digital forensics: The missing piece of the internet of things promise. Comput. Fraud Secur. **2016**(6), 5–8 (2016). https://doi.org/10.1016/s1361-3723(15)30045-2
10. A. Al-Abassi, H. Karimipour, H.H. Pajouh, A. Dehghantanha, R.M. Parizi, Industrial big data analytics: Challenges and opportunities, in *Handbook of Big Data Privacy*, ed. by K.-K. R. Choo, A. Dehghantanha, (Springer, Cham, 2020), pp. 37–61
11. M. Alaeiyan, A. Dehghantanha, T. Dargahi, M. Conti, S. Parsa, A multilabel fuzzy relevance clustering system for malware attack attribution in the edge layer of cyber-physical networks. ACM Trans. Cyber-Phys. Syst. **4**(3), 1–22 (2020)
12. M. Saharkhizan, A. Azmoodeh, A. Dehghantanha, K.-K.R. Choo, R.M. Parizi, An ensemble of deep recurrent neural networks for detecting IoT cyber attacks using network traffic. IEEE Internet Things J. **7**(9), 8852–8859 (2020). https://doi.org/10.1109/jiot.2020.2996425
13. E.M. Dovom, A. Azmoodeh, A. Dehghantanha, D.E. Newton, R.M. Parizi, H. Karimipour, Fuzzy pattern tree for edge malware detection and categorization in IoT. J. Syst. Archit. **97**, 1–7 (2019)
14. M. Kravchik, A. Shabtai, Detecting cyber attacks in industrial control systems using convolutional neural networks, in *Proceedings of the 2018 Workshop on Cyber-Physical Systems Security and Privacy*, (2018), pp. 72–83

15. K.N. Junejo, J. Goh, Behaviour-based attack detection and classification in cyber physical systems using machine learning, in *Proceedings of the 2nd ACM International Workshop on Cyber-Physical System Security*, (2016), pp. 34–43

16. J. Inoue, Y. Yamagata, Y. Chen, C.M. Poskitt, J. Sun, Anomaly detection for a water treatment system using unsupervised machine learning, in *2017 IEEE International Conference on Data Mining Workshops (ICDMW)*, (2017), pp. 1058–1065

17. M. Saharkhizan, A. Azmoodeh, H. HaddadPajouh, A. Dehghantanha, R.M. Parizi, G. Srivastava, A hybrid deep generative local metric learning method for intrusion detection, in *Handbook of Big Data Privacy*, (Springer, Cham, 2020), pp. 343–357. https://doi.org/10.1007/978-3-030-38557-6_16

18. A. Azmoodeh, A. Dehghantanha, Big data and privacy: Challenges and opportunities, in *Handbook of Big Data Privacy*, (Springer, Cham, 2020), pp. 1–5. https://doi.org/10.1007/978-3-030-38557-6_1

19. D.R. McKinnel, T. Dargahi, A. Dehghantanha, K.-K.R. Choo, A systematic literature review and meta-analysis on artificial intelligence in penetration testing and vulnerability assessment. Comput. Electr. Eng. **75**, 175–188 (2019)

20. J.C. Cabello, H. Karimipour, A.N. Jahromi, A. Dehghantanha, R.M. Parizi, Big-data and cyber- physical systems in healthcare: Challenges and opportunities, in *Handbook of Big Data Privacy*, ed. by K.-K. R. Choo, A. Dehghantanha, (Springer, Cham, 2020)

21. M. Conti, T. Dargahi, A. Dehghantanha, Cyber threat intelligence: Challenges and opportunities, in *Advances in Information Security*, (Springer, 2018), pp. 1–6. https://doi.org/10.1007/978-3-319-73951-9_1

22. StratoEnergetics, Slaughterbots (2017)

23. A. Yazdinejad, R.M. Parizi, A. Dehghantanha, G. Srivastava, S. Mohan, A.M. Rababah, Cost optimization of secure routing with untrusted devices in software defined networking. J. Parallel Distrib. Comput. **143**, 36–46 (2020)

24. A. Yazdinejad, R.M. Parizi, A. Dehghantanha, K.-K.R. Choo, Blockchain-enabled authentication handover with efficient privacy protection in SDN-based 5G networks. IEEE Trans. Netw. Sci. Eng., 1–1 (2019). https://doi.org/10.1109/tnse.2019.2937481

25. A. Azmoodeh, A. Dehghantanha, M. Conti, K.-K.R. Choo, Detecting crypto-ransomware in IoT networks based on energy consumption footprint. J. Ambient. Intell. Humaniz. Comput. **9**(4), 1141–1152 (2018)

26. A. Azmoodeh, A. Dehghantanha, K.-K.R. Choo, Robust malware detection for internet of (battlefield) things devices using deep eigenspace learning. IEEE Trans. Sustain. Comput. **4**(1), 88–95 (2018)

27. H. Haddadpajouh, A. Azmoodeh, A. Dehghantanha, R.M. Parizi, MVFCC: A multi-view fuzzy consensus clustering model for malware threat attribution. IEEE Access **8**, 139188–139198 (2020)

28. H. HaddadPajouh, A. Dehghantanha, R. Khayami, K.-K.R. Choo, A deep recurrent neural network based approach for internet of things malware threat hunting. Futur. Gener. Comput. Syst. **85**, 88–96 (2018). https://doi.org/10.1016/j.future.2018.03.007

29. A.N. Jahromi, S. Hashemi, A. Dehghantanha, R.M. Parizi, K.-K.R. Choo, An enhanced stacked LSTM method with no random initialization for malware threat hunting in safety and time-critical systems. IEEE Trans. Emerg. Top. Comput. Intell. **4**(5), 630–640 (2020). https://doi.org/10.1109/tetci.2019.2910243

30. S. Homayoun et al., Deep dive into ransomware threat hunting and intelligence at fog layer. Futur. Gener. Comput. Syst. **90**(Jan 19), 94–104 (2018)

31. S. Homayoun, A. Dehghantanha, M. Ahmadzadeh, S. Hashemi, R. Khayami, Know abnormal, find evil: Frequent pattern mining for ransomware threat hunting and intelligence. IEEE Trans. Emerg. Top. Comput. (2017). https://doi.org/10.1109/TETC.2017.2756908

32. S. Homayoun et al., DRTHIS: Deep ransomware threat hunting and intelligence system at the fog layer. Futur. Gener. Comput. Syst. **90**, 94–104 (2019). https://doi.org/10.1016/j.future.2018.07.045

33. A. Aminnezhad, A. Dehghantanha, M.T. Abdullah, A survey on privacy issues in digital forensics. Int. J. Cyber-Secur. Digit. Forensics **1**(4), 311–324 (2012)
34. J. Goh, S. Adepu, K.N. Junejo, A. Mathur, A dataset to support research in the design of secure water treatment systems, in *International Conference on Critical Information Infrastructures Security*, (2016), pp. 88–99
35. A. Yazdinejad, A. Bohlooli, K. Jamshidi, Performance improvement and hardware implementation of open flow switch using FPGA, in *IEEE 5th Conference on Knowledge Based Engineering and Innovation, KBEI 2019*, (2019), pp. 515–520
36. A. Yazdinejad, R.M. Parizi, A. Dehghantanha, H. Karimipour, G. Srivastava, M. Aledhari, Enabling drones in the internet of things with decentralized blockchain-based security. IEEE Internet Things J., 1 (2020). https://doi.org/10.1109/jiot.2020.3015382
37. A. Singh, K. Click, R.M. Parizi, Q. Zhang, A. Dehghantanha, K.-K.R. Choo, Sidechain technologies in blockchain networks: An examination and state-of-the-art review. J. Netw. Comput. Appl. **149**, 102471 (2020). https://doi.org/10.1016/j.jnca.2019.102471
38. A. Yazdinejad, R.M. Parizi, A. Dehghantanha, Q. Zhang, K.-K.R. Choo, An energy-efficient SDN controller architecture for IoT networks with blockchain-based security. IEEE Trans. Serv. Comput. **13**(4), 625–638 (2020)
39. D. Połap, G. Srivastava, A. Jolfaei, R.M. Parizi, Blockchain technology and neural networks for the internet of medical things, in *IEEE INFOCOM 2020 – IEEE Conference on Computer Communications Workshops (INFOCOM WKSHPS)*, (2020), pp. 508–513. https://doi.org/10.1109/INFOCOMWKSHPS50562.2020.9162735
40. A. Yazdinejad, G. Srivastava, R.M. Parizi, A. Dehghantanha, K.-K.R. Choo, M. Aledhari, Decentralized authentication of distributed patients in hospital networks using blockchain. IEEE J. Biomed. Health Inform. **24**(8), 2146–2156 (2020)
41. Q. Chen, G. Srivastava, R.M. Parizi, M. Aloqaily, I. Al Ridhawi, An incentive-aware blockchain-based solution for internet of fake media things. Inf. Process. Manag., 102370 (2020). https://doi.org/10.1016/j.ipm.2020.102370
42. A. Yazdinejad, R.M. Parizi, A. Bohlooli, A. Dehghantanha, K.-K.R. Choo, A high-performance framework for a network programmable packet processor using P4 and FPGA. J. Netw. Comput. Appl. **156**, 102564 (2020)
43. R.M. Parizi, S. Homayoun, A. Yazdinejad, A. Dehghantanha, K.-K.R. Choo, Integrating privacy enhancing techniques into blockchains using sidechains, in *Proceedings of the 32nd IEEE Canadian Conference on Electrical and Computer Engineering (CCECE 2019)*, (2019). https://doi.org/10.1109/CCECE.2019.8861821
44. A. Yazdinejad, R.M. Parizi, G. Srivastava, A. Dehghantanha, K.-K.R. Choo, Energy efficient decentralized authentication in internet of underwater things using blockchain, in *2019 IEEE Globecom Workshops (GC Wkshps)*, (2019), pp. 1–6
45. J. Sakhnini, *Security of Smart Cyber-Physical Grids: A Deep Learning Approach* (2020), p. 83
46. World Health Organization et al., in *HAMIDS | Proceedings of the 2nd ACM Workshop on Cyber-Physical Systems Security and Privacy,* http://10.0.4.121/2994487.2994492?casa_token=fzc-QNOcjJkAAAAA:iKofJD9cHqHxMQjOxse0v8N4Au0fAwilQzYXDm0MO4a XMQHng4p3NHbqHNFgnwN8AIQNI6T2K5G (acc *Osteoarthr. Cartil*)
47. S. Singh, S. Silakari, An ensemble approach for cyber attack detection system: A generic framework, in *2013 14th ACIS International Conference on Software Engineering, Artificial Intelligence, Networking and Parallel/Distributed Computing*, (2013), pp. 79–84
48. M.T. Khorshed, N.A. Sharma, A.V. Dutt, A.B.M.S. Ali, Y. Xiang, Real time cyber attack analysis on Hadoop ecosystem using machine learning algorithms, in *2015 2nd Asia-Pacific World Congress on Computer Science and Engineering (APWC on CSE)*, (2015), pp. 1–7
49. B.W. Masduki, K. Ramli, F.A. Saputra, D. Sugiarto, Study on implementation of machine learning methods combination for improving attacks detection accuracy on Intrusion Detection System (IDS), in *2015 International Conference on Quality in Research (QiR)*, (2015), pp. 56–64

50. F. Zhang, H.A.D.E. Kodituwakku, J.W. Hines, J. Coble, Multilayer data-driven cyber-attack detection system for industrial control systems based on network, system, and process data. IEEE Trans. Ind. Inform. **15**(7), 4362–4369 (2019)
51. A.P. Bradley, The use of the area under the ROC curve in the evaluation of machine learning algorithms. Pattern Recogn. **30**(7), 1145–1159 (1997)

Evaluation of Scalable Fair Clustering Machine Learning Methods for Threat Hunting in Cyber-Physical Systems

Dilip Sahoo and Aaruni Upadhyay

1 Introduction

Recent years has witnessed a proliferation of using computerized system for majority aspects of our today's life [1–5], which encouraged cybercriminals to attack these system by desining sophisticated attack patterns. The safety of our society and infrastructure depends on keeping our mission-critical systems such as Water distribution safe from cyber-attacks [6–10]. Many such systems work in tandem with the Internet of Things (IoT) systems and other cyber-physical systems that are susceptible to attacks by hostile nations and other non-state actors [11–15]. Machine learning is increasingly being used in designing systems that can detect such attacks through clustering which is an unsupervised machine learning technique [16–19].

The behavior of machine learning systems is dependent on the training data which may contain biases which may in return, result in the bias being reflected in the outcome [20]. This problem was highlighted by Chierichetti in [21] where they argue that the biases may still indirectly appear in results even if unprotected attributes (such as a person's height) are used for making decisions instead of protected ones such as race and gender. This could happen because of the hidden correlations that may exist between protected and unprotected attributes, for example, average height (unprotected) is related to gender (protected) and can be exploited as a proxy for discrimination.

D. Sahoo
Cyber Science Lab, University of Guelph, Guelph, ON, Canada
e-mail: dilip@cybersciencelab.org

A. Upadhyay (✉)
MCTI, University of Guelph, Guelph, ON, Canada
e-mail: aupadhya@uoguelph.ca

© Springer Nature Switzerland AG 2021
K.-K. R. Choo, A. Dehghantanha (eds.), *Handbook of Big Data Analytics and Forensics*, https://doi.org/10.1007/978-3-030-74753-4_10

The established approach followed by the machine learning researchers to solve this problem can be traced back to the US Supreme Court case Griggs v. Duke Power Co. [22] that resulted in the emergence of the concept of adverse impact. Adverse impact occurs when a practice negatively and disproportionately affects a protected group regardless if it was indirectly or unintentionally. The "80% rule" was adopted by the researchers as a generally accepted way to measure adverse impacts which states that an adverse impact has occurred if "the selection rate for a certain group is less than 80 percent of that of the group with the highest selection rate" [23].

Chierichetti applied this notion of fairness to clustering by introducing the use of fairlets that groups together the datapoints while preserving the fairness objective. These fairlets are then combined to form clusters by using existing k-median algorithms. This way, fair clustering reduces biases by placing constraints on the clusters so that the probability of a class of input data points being present in a cluster, is strictly greater than zero. However fair clustering achieved using this method has a super-quadratic runtime. The paper we are basing our research on [24] presents a new implementation of this fair clustering method that runs in near-linear time and therefore offers performance that scales with the input size.

To formally outline the problem, we must first define fair clustering and we will use the same definition as our base paper. Consider n number of points P from the training dataset such that each point belongs to one of two types: $T1$ and $T2$. In a practical application, these classes can correspond to any legally protected attribute such as gender where $T1$: *Male* and $T2$: *Female*. Let's define the Balance of a subset S such that $S \subseteq P$, assuming S_{T1} and S_{T2} represent subsets of $T1$ and $T2$ in the set S.

$$\text{Balance}(S) = \min \left\{ \left| \frac{S_{T1}}{S_{T2}} \right|, \left| \frac{S_{T2}}{S_{T1}} \right| \right\}$$

If we assume $T1 < T2$, then the clustering of P performed over $(T1, T2)$ would be defined as fair if for all clusters C:

$$\text{Balance}(C) \geq \left(\frac{T1}{T2} \right)$$

A formal definition of k-median fair clustering can now be stated as the division of input point set P into k clusters such that the sum of distances of each point $p \in P$ to the center of their cluster is minimum AND all clusters have a balance of at least $\left(\frac{T1}{T2} \right)$.

Our contribution through this research is to run the k-median fair clustering implementation of original authors [24] on 4 new Cybersecurity related datasets and evaluate the performance and accuracy of our results. We demonstrate that our algorithm runtime is near-linear which is the same as expected in the original paper and so we show that the algorithm is scalable also for much bigger datasets like SWaT. The datasets used in our experiment are referenced in [25–27].

Section 2 of this paper contains a Literature Review of the recent work done in fair clustering. Section 3 details the methodology of our experiment and is followed by a discussion of our Experimentation and Results in Sect. 4. A comparison of our findings with that of the base paper is presented in Sect. 5. Our concluding statements and avenue for future work are presented in Sect. 6.

2 Related Work

Our dependence on critical systems like electricity and water distribution systems makes them a very lucrative target for our adversaries. Similarly, IoT networks that are frequently integrated with such systems are a frequent target for attack and are also used as a vector for further malware spread and launching DDoS attacks [28–30]. Machine learning researchers have been actively exploring ways to detect attacks on these crucial systems by utilizing machine learning techniques [31]. and [32] are an example of the use of machine learning techniques in the detection of threats in IoT and Water distribution systems respectively.

Due to an ever increasing adoption of IoT systems [33–38], malware detection in these devices [39–43] has become a topic of great interest among cybersecurity researchers [44–46] Authors in [47] highlight the new paradigm of edge computing in IoT networks and demonstrate the use of fuzzy and fast fuzzy pattern tree methods for detecting IoT malware. Authors in [48] present an interesting approach to detect the presence of ransomware in IoT networks by monitoring the power consumption patterns of IoT devices. Their machine learning based approach was successfully able to classify ransomware from non-malicious applications and produced a better accuracy and precision rate than K-Nearest Neighbors, Support Vector Machine, Neural Network and Random Forest methods [49]. presents a approach for detecting intrusion in IoT networks based on two-layer dimension reduction and two-tier classification module to detect User-To-Root (U2R) and Remote-To-Local (R2L) attacks. The authors use the NSL-KDD dataset and demonstrate that their approach performs better than earlier models designed to detect R2L and U2R attacks. Attackers often employ the use of code level polymorphism to evade any opcode based malware detection algorithms. Authors in [50] demonstrate the use of sequential pattern mining approach to select best features to train KNN, SVM, AdaBoost and other machine learning models and are able to detect IoT malware with polymorphed code to escape detection.

One pitfall of using machine learning can be the appearance of bias in the output if we are not careful. The authors in [51] present the case of a medical center that used an algorithm, that was used to screen patients in an intensive care program, to be racially biased against black patients. The algorithm was found to be functioning correctly, but the bias was inadvertently introduced because it wrongly established that black patients are healthier because they spend less on healthcare. Bias in real-world computing applications can have serious ethical implications as highlighted in [52]. Authors argue that any attempt at fairness, even if it is not 100% effective, should be incorporated in our algorithms instead of waiting for a perfect fair algorithm to emerge. The goal of the fairness algorithms should not be to have a perfect solution to the fairness problem but instead to maximize the common good by achieving whatever fairness is attainable today.

The notion of fair clustering was first introduced in [21] who articulated the implementation of fair clustering for both k-center and k-median objectives. They introduced the idea of division of pointset into smaller minimal subsets (called *fairlets*) that fulfill the fairness criteria while meeting the clustering objective. They used 3 datasets (*Diabetes, Bank, and Census*) to evaluate their algorithm and compare the performance and fairness of their approach to the classical k-center and k-median algorithms. The results successfully demonstrated that traditional k-center and k-median algorithms produced unfair clusters as compared to their fair algorithms. However, their fair algorithm was computationally harder than the traditional algorithms.

Several researchers have done subsequent work based on the original work in [21]. Authors in [53] study the problem of low-cost fair clustering where the data points can belong to multiple protected classes. Their implementation allows for placing upper and lower bounds for any class in a cluster while maintaining fairness for data points that may even span multiple protected classes [54]. looks at the effects of using *fairlet* based approach in fair clustering and raises the important concern of scalability of that algorithm. They propose the use of the concept of *coresets* that is tailored for use in fair clustering problem to provide their own algorithms for fair k-means clustering as an improvement on algorithms in [21]. They empirically demonstrate how *coresets* enable fair clustering algorithms and improve output quality by using better albeit slower algorithms.

Several works have also emerged highlighting the usage of fair algorithms in real applications [55]. uses the fair design for preserving privacy by adding a constraint so that a cluster will be formed only if a lower bound on the number of points is achieved to preserve anonymity. Authors in [56] apply the notion of fairness to address the allocation problem where we want to distribute a limited resource to be distributed across different clusters without bias. For example, the allocation of housing loans based on creditworthiness across groups without a prejudice based on race. Authors in [57] propose fairness preserving algorithms that can produce a summary of texts (e.g. blog posts) in a way so that it fairly represents the opinion of different social groups.

3   Methodology

This section describes the detailed steps taken to evaluate the fair clustering Machine Learning (ML) model using four different datasets. Each of the four datasets contained data labeled as 'normal/good ware' and 'attack/malware' which were collected from the respective testbeds under normal and attack scenarios. The raw datasets were first pre-processed and then feature selection and extraction were applied to reduce overall dimensionality. Finally, the scalable fair clustering algorithm [26] was adopted as the ML classifier for the experiment and the model

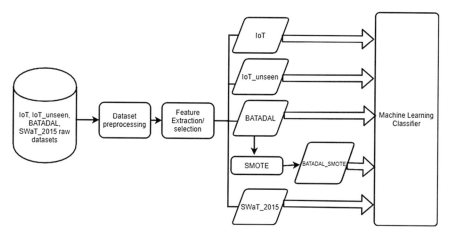

Fig. 1 Experiment workflow structure

performance was measured based on cost, runtime, and accuracy. Figure 1 illustrates the experiment workflow structure.

3.1 Datasets Preprocessing

The preprocessing of each dataset was done in separate ways based on the type and nature of the raw data samples. The detailed preprocessing approach for them is described below.

3.1.1 IoT Dataset

The IoT dataset [27] had a total of 614 text files in the dataset among which 362 files were labeled as 'good ware' and 252 as 'malware'. All the text files contained the opcode data collected from the testbed. These individual text files were processed using Term Frequency- Inverse Document Frequency (TF-IDF) and a TF-IDF score was assigned to each of the opcodes. After, assigning the TF-IDF score the opcodes were treated as feature columns for the modified dataset with the column header as the opcode names and rows as the TF-IDF score assigned to them. Each row created corresponded to one of the text files in the raw data samples and finally, we obtained a CSV file having a total of 236 feature columns and 512 rows after removing anomalies. In the obtained CSV file, there were 271 'good ware' and 281 'malware' data samples and were labeled as 0 and 1, respectively. Hence, we got our final CSV file with all data samples in a numerical format.

3.1.2 IoT_unseen Dataset

The IoT_unseen dataset [27] had a similar raw data format as the IoT dataset described in the previous section, except there were differences in the number and type of data samples. This dataset was used mainly for the ML model accuracy evaluation purpose. It is worth mentioning that this dataset was not used in all stages of experiments performed for the main IoT dataset and the rest of the datasets because this dataset is equivalent to the IoT dataset in nature. This dataset had 51 opcode-based text files and all of them were labeled as 'malware'. A similar TF-IDF based processing was done again to obtain a CSV file with 51 rows against each of the raw text files and 666 feature columns.

3.1.3 BATADAL

It is an industrial control system (ICS) dataset [26] collected during 'Normal' and 'Attack' scenarios from different sensors of the testbed. The dataset contains 12,446 'Normal' and 492 'Attack' data samples. During data preprocessing the .npy file was converted to CSV using 'numpy' library (numpy.org) and all features were converted to numerical data.

3.1.4 SWaT_2015

SWaT_2015 data was collected from the testbed of a six-stage secure water treatment system under 'Normal' and 'Attack' scenarios [25]. The dataset contains 1,387,095 'Normal' data samples and 54,621 'Attack' samples. During preprocessing of this dataset, a similar approach as 'BATADAL' was used and the final CSV file was obtained.

3.2 Feature Selection and Feature Extraction

For feature selection and extraction, we implemented SelectKBest with Chi^2 and ExtraTreeClassifier feature scoring method from scikit-learn (scikit-learn.org). The fair clustering ML model [58] adopted for our experiment requires at least one feature column as the sensitive attribute with categorical data. Hence, One of the feature columns, from every four datasets, was decided to be taken as a sensitive attribute and all the values present under the sensitive attribute were converted to categorical data i.e. either 1 or 0. It is worth noting that the balance parameters derived from the ratio of categorical values(1 or 0) in the sensitive attribute are used to create the target fairlet decomposition clusters. Table 1 shows the sensitive attribute name and number of other features selected for the ML model accuracy evaluation for each dataset during the experiment.

Table 1 Sensitive attribute and number of selected features for the datasets

Dataset	Total number of data points	Sensitive attribute	No. of selected features
IoT	512	stmgeia	12
IoT_unseen	51	bnd	665
BATADAL	12,938	P_J422_code	9
BATADAL_SMOTE	24,892	P_J422_code	9
SWaT_2015	1,441,715	UV401	5

3.2.1 Upsampling Using Synthetic Minority Over-Sampling Technique (SMOTE)

In the 'BATADAL' dataset, we observed a huge variance in the number of data points between 'Normal' and 'Attack' families. To ensure that our fair clustering ML model is not biased towards the majority class, we used SMOTE technique to balance the majority and minority class. The new dataset after upsampling held 12,446 data points for each of the 'Normal' and 'Attack' class. The new SMOTE enhanced BATADAL dataset is referred to as 'BATADAL_SMOTE'. Accuracy evaluation for the 'BATADAL_SMOTE' dataset was done separately to compare it with the accuracy of the original BATADAL dataset during the experiments.

3.3   Fair Clustering ML Model Implementation

In this phase, the fair clustering ML model was fed with the extracted features of all four datasets. All the experiments were conducted in a Windows 10 virtual machine environment with 2.21 GHz 64-bit intel i7 processor and 4GB RAM. Jupyter Notebook was used with python 3.6.5 and MATLAB engine.

4   Experiments and Results

This section describes the details of three individual experiments conducted and highlights the results. The evaluation measures used to assess the results are described first in Sect. 4.1 and then the details of conducted experiments and their results are discussed in Sects. 4.2, 4.3 and 4.4. The first experiment has been conducted to evaluate the performance of the ML model based on the overall cost and runtime. In the second experiment, the model accuracy is evaluated for every dataset. A separate experiment is done to test the effectiveness of the fair clustering algorithm by comparing its results with the results of a relevant normal clustering algorithm described in Sect. 4.4.

4.1 Evaluation Measures

In our first experiment described in Sect. 4.2, evaluation of the near-linear behavior and performance of the fair clustering algorithm is done by measuring the fairlet decomposition cost, fair clustering cost, and runtime. The fairlet decomposition cost denotes the distance between the points and their cluster centroids. Fair clustering cost is the total algorithm cost and the runtime states the time taken in seconds to run the clustering on a certain number of data points.

For the second and third experiments, we used commonly used machine learning matrices which are discussed below. A Confusion matrix represents the summary of all the predicted results of an algorithm in terms of the number of True Positives (TP), True Negatives (TN), False Positives (FP) and False Negatives (FN). Indicators used to assess the results of the fair clustering algorithm in our experiment are derived from the confusion matrix. Calculation of 'Precision' and 'Recall' is based on the binary label i.e. 'Normal' and 'Attack'. 'F1-score' combines precision and recall and to provide the clustering performance. 'Accuracy' denotes how accurately the clustering algorithm detects the binary classes i.e. 'Normal' or 'Attack'.

TP: Normal observation is predicted as normal
TN: Attack observation is predicated as an attack
FP: Attack observation is predicted as normal
FN: Normal observation is predicated as an attack

$$\textbf{Accuracy} = \frac{TP + TN}{TP + TN + FP + FN}$$

$$\textbf{Precision} = \frac{TP}{TP + FP}$$

$$\textbf{Recall} = \frac{TP}{TP + FN}$$

$$\textbf{F1} - \textbf{score} = 2 * \left(\frac{Precision * Recall}{Precision + Recall} \right)$$

4.1.1 Accuracy Evaluation for Two Clusters and Multicluster Clustering

To determine the performance of the fair clustering algorithm output results in terms of 'Accuracy' using two clusters(k = 2), it is assumed that all data points predicted under one cluster(cluster-1) are 'Normal' and the other cluster (cluster-2) are of 'Attack' type. Then, it is calculated how many data points in cluster-1 are indeed 'Normal' and cluster-2 are indeed 'Attack' from the known labels. This resulted in the confusion matrix construction where the diagonal elements represented TPs and TNs respectively along with the off-diagonal elements as FPs and FNs.

In contrast to clustering with $k = 2$, we have used a different approach for multiple clustering ($k = n$, where the value of n is 4,6,10,15 and 20). In the case of multiple clustering, we assume each of the data points in a cluster as 'Normal' if the actual labels for most of the data points are of 'Normal' type. Similarly, in a cluster, if the actual labels of majority data points are of 'Attack' type then every data point belongs to that cluster is assumed as 'Attack' and confusion matrix was constructed. In the multicluster clustering, the final accuracy is the average accuracy of each cluster.

4.2   Experiment-I and Results

In Experiment-I, all datasets described in Sect. 3 except 'IoT_unseen' and 'BATADAL_SMOTE' were fed into the fair clustering algorithm. The input dimension and balance parameters for the dataset during the experiment are mentioned in Table 2. We did not use all features extracted for Experiment-II accuracy evaluation in this phase of the experiment. The performance of the algorithm was measured in terms of cost and runtime with a different number of data samples and dimensions for all datasets. The results in Table 2 shows that the algorithm performance varies depending on the overall dimension and size of the dataset. It can be noticed that the IoT dataset having a high dimension with only 500 data samples took 20 seconds to complete the whole clustering process. In contrast, the SWaT_2015 dataset with 40 times more data samples than the IoT dataset took little more than twice the time needed for IoT.

Table 2 Performance evaluation of each of the datasets for a specific number of data samples in terms of runtime and cost

Dataset	Dimension	Balance	No of clusters $k = 20$			
			Fairlet decomposition cost	Fair clustering cost	Fairlet decomposition time (in sec)	Total time (in sec)
IoT (500 sample)	11	0.5	50.8	56.4	0.29	20
BATADAL (1000 sample)	6	0.33	142	205	8.4	37
SWaT_2015 (20,000 sample)	2	0.02	17,809	39,209	10	41

In addition to the performance evaluation of the fair clustering algorithm, we also checked the scalability of the fair clustering algorithm in terms of runtime for a range of data points. All three datasets were fed into the fair clustering algorithm with a different number of data samples more than once and the time taken to complete the process for each sample was recorded. We then plotted the graphs between the number of data samples Vs runtime to investigate whether the algorithm scales in linear time. Figures 2, 3 and 4 show the graphs between the number of data points Vs runtime for the three datasets. The results show that the plotted graphs for SWaT_2015 and BATADAL datasets are almost a straight line with a slight deviation for the IoT dataset.

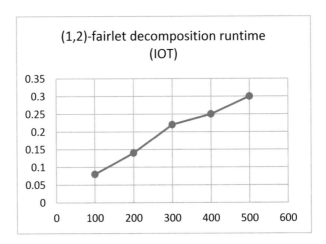

Fig. 2 No of datapoints Vs Runtime (in sec) graph for IOT dataset

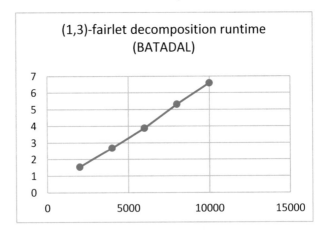

Fig. 3 No of datapoints Vs Runtime (in sec) graph for BATADAL dataset

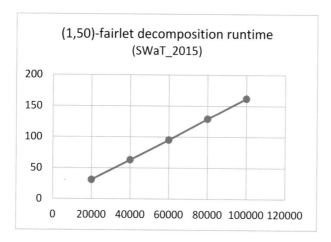

Fig. 4 No of datapoints Vs Runtime (in sec) graph for SWaT_2015 dataset

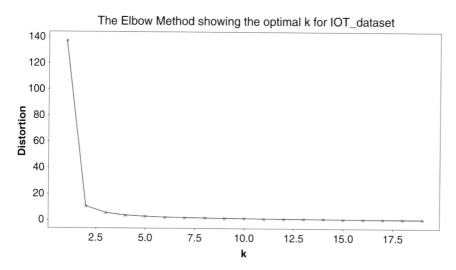

Fig. 5 Elbow method showing optimum k for IoT dataset

4.3 Experiment-II and Results

In experiment-II, all five datasets described in Table 1 are fed into the fair clustering algorithm and the desired numbers of clusters were obtained as output. We used the elbow method from scikit-learn to predict the optimum number of cluster(k) for each dataset.

Figures 5, 6, 7 and 8 illustrate the optimum cluster value(k) for IoT, IoT_unseen, BATADAL, and SWaT_2015 datasets respectively using the elbow method.

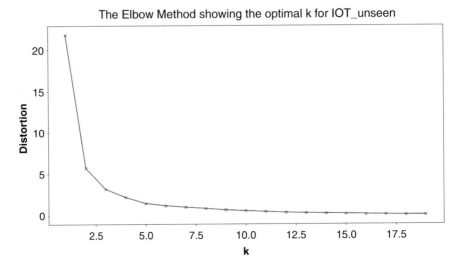

Fig. 6 Elbow method showing optimum k for IoT_unseen dataset

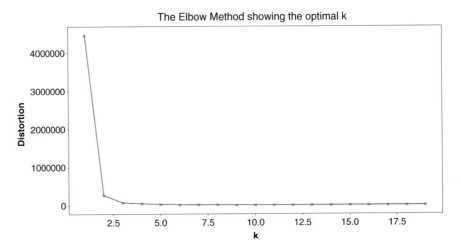

Fig. 7 Elbow method showing optimum k for BATADAL dataset

After having an idea about the optimum cluster value for each dataset, we decided to run each dataset with different values of 'k' i.e. k = 2, 4, 6, 10, 15 and 20. The performance of fair clustering algorithm is measured in terms of 'Accuracy', 'Precision', 'Recall', and 'F1-score' based on the method discussed in Sect. 4.1.1. The results obtained from the ML model for each dataset are shown in Table 3. IoT dataset achieved the highest accuracy of 99% with two clusters (k = 2) and the accuracy remained slightly less with other values of cluster numbers(k). IoT_unseen obtained an accuracy of 80% for 'k-value' as 2 and it

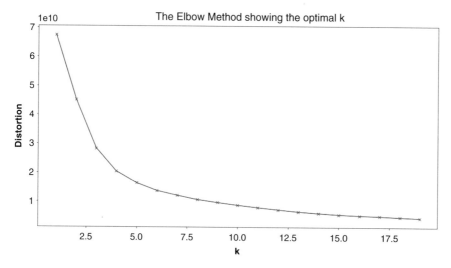

Fig. 8 Elbow method showing optimum k for SWaT_2015 dataset

could obtain 100% accuracy when 'k-value' was 4,6,10,15 and 20. BATADAL and BATADAL_SMOTE dataset shown a significant difference in their result. BATADAL dataset achieved maximum accuracy of 97% compared to the maximum accuracy of 67% for the BATADAL_SMOTE dataset. The reason behind such difference in results of BATADAL and BATADAL_SMOTE is attributed to the BATADAL dataset biased with majority class. The SWaT_2015 dataset reported a maximum accuracy of 97% with 'k-value' as 6. Overall, the fair clustering algorithm predicted the best results for IoT and IoT_unseen datasets among other datasets.

4.4 Experiment-III and Results

Experiment-III is conducted to evaluate the suggested fair clustering algorithm performance in terms of 'Accuracy' by comparing it with the equivalent normal clustering algorithm. The exact datasets taken for Experiment-II(with fair clustering step) were fed into 'kmedoids' clustering algorithm [59] from MATLAB engine without combining the fairlet decomposition step. The 'Accuracy' for all datasets without fair clustering is recorded and then compared with the respective results from Experiment-II. Table 4 shows the comparison of the results between fair clustering and normal clustering with 'k-value' as 2.

Table 3 Results obtained from fair clustering algorithm

Data set	No of clusters(k)	Accuracy (%)	Precision (%)	Recall (%)	F1-score (%)
IoT	2	99	99	99	99
	4	97	97	97	97
	6	97	97	97	97
	10	98	98	98	98
	15	97	98	97	98
	20	99	99	99	99
IoT_unseen	2	80	100	80	89
	4	100	100	100	100
	6	100	100	100	100
	10	100	100	100	100
	15	100	100	100	100
	20	100	100	100	100
BATADAL	2	81	93	81	86
	4	96	97	96	95
	6	97	97	97	95
	10	97	96	97	96
	15	97	97	97	96
	20	97	97	97	96
BATADAL_SMOTE	2	59	64	59	57
	4	62	65	62	61
	6	62	62	62	62
	10	66	69	66	65
	15	66	69	66	66
	20	67	67	67	66
SWaT_2015	2	74	86	74	79
	4	93	93	93	93
	6	97	96	97	96
	10	95	95	95	95
	15	86	85	86	84
	20	96	96	96	96

Table 4 Results comparison between fair clustering and normal clustering

Data set	Accuracy (%) with k = 2	
	Fair_clustering	Without Fair_clustering
IoT	99	92
IoT_unseen	80	61
BATADAL	81	82
BATADAL_SMOTE	59	57
SWaT_2015	74	74

5   Results Comparison

The research done in [58] has evaluated the performance of the suggested fair clustering algorithm in terms of cost and runtime. However, the authors did not provide the details of the accuracy they achieved. The results of our experiment show that the fair clustering algorithm achieved better accuracy than the normal clustering algorithm for every dataset except BATADAL and SWaT_2015 where both performed almost equally.

6   Conclusion and Future Work

We successfully implemented a scalable fair clustering machine learning algorithm and evaluated the model with different cyber-physical datasets. Our results show that the model scales linearly with the number of input data points even for a large dataset like SWaT_2015. The unsupervised fair clustering algorithm was run with different cluster values ($k = 2,4,6,10,15$ and 20) as well as different input balance parameters (mentioned in Table 2) which provided us with a comprehensive view of our results. Our model achieved an accuracy of 99% with a precision of 99% and recall of 99% for IoT dataset which was best among all other datasets. While evaluating the IoT_unseen dataset, the model accomplished classification with 100% accuracy. Although we tried to address the majority class biasing issue in BATADAL dataset by creating synthetic data samples, the accuracy of BATADAL_SMOTE remained relatively lower at 67%. Overall, all our datasets exhibited high accuracy of above 95% except BATADAL_SMOTE.

For future work, the fair clustering model implemented in this paper could be tested on other different cyber-physical systems. Due to the resource limitation of our research environment, we could only feed up to 20,000 samples for accuracy evaluation, but this limit can be increased to improve the model prediction capabilities. While up-sampling datasets with artificial data provided a good method for balancing the BATADAL dataset, it is not as effective as original data collected from a real-world scenario. We expect that datasets collected from real-world systems with higher data quality will improve the resulting outcome of our model.

References

1. S. Nakhodchi, A. Dehghantanha, H. Karimipour, Privacy and security in smart and precision farming: A bibliometric analysis, in *Handbook of Big Data Privacy*, (Springer, Cham, 2020), pp. 305–318
2. S. Walker-Roberts, M. Hammoudeh, A. Dehghantanha, A systematic review of the availability and efficacy of countermeasures to internal threats in healthcare critical infrastructure. IEEE Access **6**, 25167–25177 (Mar. 2018). https://doi.org/10.1109/ACCESS.2018.2817560

3. H.M. Rouzbahani, H. Karimipour, A. Dehghantanha, R.M. Parizi, Blockchain applications in power systems: A bibliometric analysis, in *Blockchain Cybersecurity, Trust and Privacy*, ed. by K.-K. R. Choo, A. Dehghantanha, R. M. Parizi, vol. 79, (Springer, Cham)
4. A. Yazdinejad, R.M. Parizi, A. Dehghantanha, H. Karimipour, G. Srivastava, M. Aledhari, Enabling drones in the internet of things with decentralized Blockchain-based security, IEEE Internet Things J., 1 (2020). https://doi.org/10.1109/jiot.2020.3015382
5. K. Bolouri, A. Azmoodeh, A. Dehghantanha, M. Firouzmand, Internet of things camera identification algorithm based on sensor pattern noise using color filter array and wavelet transform, in *Handbook of Big Data and IoT Security*, (Springer, Cham, 2019), pp. 211–223. https://doi.org/10.1007/978-3-030-10543-3_9
6. S. Watson, A. Dehghantanha, Digital forensics: The missing piece of the internet of things promise. Comput. Fraud Secur. **2016**(6), 5–8 (2016). https://doi.org/10.1016/s1361-3723(15)30045-2
7. F. Daryabar, A. Dehghantanha, N. I. Udzir, N. F. B. M. Sani, S. Bin Shamsuddin, Towards secure model for SCADA systems. IEEE Xplore (2012, June 1). https://doi.org/10.1109/CyberSec.2012.6246111
8. A. Azmoodeh, A. Dehghantanha, K.-K.R. Choo, Big data and internet of things security and forensics: Challenges and opportunities, in *Handbook of Big Data and IoT Security*, (Springer, Cham, 2019), pp. 1–4. https://doi.org/10.1007/978-3-030-10543-3_1
9. M. Conti, T. Dargahi, A. Dehghantanha, Cyber threat intelligence: Challenges and opportunities, in *Advances in Information Security*, (Springer, Cham, 2018), pp. 1–6. https://doi.org/10.1007/978-3-319-73951-9_1
10. S. Grooby, T. Dargahi, A. Dehghantanha, Protecting IoT and ICS platforms against advanced persistent threat actors: Analysis of APT1, silent chollima and molerats, in *Handbook of Big Data and IoT Security*, (Springer, Cham, 2019), pp. 225–255
11. H. Karimipour, V. Dinavahi, Robust massively parallel dynamic state estimation of power systems against cyber-attack. IEEE Access **6**, 2984–2995 (2017)
12. P.N. Bahrami, A. Dehghantanha, T. Dargahi, R.M. Parizi, K.-K.R. Choo, H.H.S. Javadi, Cyber kill chain-based taxonomy of advanced persistent threat actors: Analogy of tactics, techniques, and procedures. J. Inf. Process. Syst. **15**(4), 865–889 (2019)
13. H. Haddadpajouh, A. Azmoodeh, A. Dehghantanha, R.M. Parizi, MVFCC: A multi-view fuzzy consensus clustering model for malware threat attribution. IEEE Access **8**, 139188–139198 (2020)
14. H. Darabian et al., A multiview learning method for malware threat hunting: Windows, IoT and android as case studies. World Wide Web **23**(2), 1241–1260 (2020)
15. A. Yazdinejad, R. M. Parizi, A. Dehghantanha, K.-K. R. Choo, Blockchain-enabled authentication handover with efficient privacy protection in SDN-based 5G networks, *IEEE Trans. Netw. Sci. Eng.*, pp. 1–1 (2020). https://doi.org/10.1109/TNSE.2019.2937481
16. J. Sakhnini, H. Karimipour, A. Dehghantanha, R. M. Parizi, G. Srivastava, Security aspects of Internet of Things aided smart grids: A bibliometric survey, in *Internet of Things*, (2019), p. 100111. https://doi.org/10.1016/j.iot.2019.100111
17. A. Yazdinejad, R.M. Parizi, A. Dehghantanha, K.-K.R. Choo, P4-to-blockchain: A secure blockchain-enabled packet parser for software defined networking. Comput. Secur. **88**, 101629 (2020). https://doi.org/10.1016/j.cose.2019.101629
18. A.N. Jahromi et al., An improved two-hidden-layer extreme learning machine for malware hunting. Comput. Secur. **89**, 101655 (2020)
19. A. Azmoodeh, A. Dehghantanha, R.M. Parizi, S. Hashemi, B. Gharabaghi, G. Srivastava, Active spectral botnet detection based on eigenvalue weighting, in *Handbook of Big Data Privacy*, (Springer, Cham, 2020), pp. 385–397. https://doi.org/10.1007/978-3-030-38557-6_19
20. H. Karimipour, A. Dehghantanha, R.M. Parizi, K.-K.R. Choo, H. Leung, A deep and scalable unsupervised machine learning system for cyber-attack detection in large-scale smart grids. IEEE Access **7**, 80778–80788 (2019)

21. F. Chierichetti, R. Kumar, S. Lattanzi, S. Vassilvitskii, Fair clustering through fairlets, in *Advances in Neural Information Processing Systems*, (MIT Press, Cambridge, 2017), pp. 5029–5037

22. World Health Organization et al., *U.S. Reports: Griggs v. Duke Power Co., 401 U.S. 424* (Library of Congress, Washington, DC, 1971) https://www.loc.gov/item/usrep401424/. Accessed 18 Sep 2020. *Osteoarthr. Cartil*

23. N. Mondragon, in *What is Adverse Impact? And Why Measuring It Matters* (2018, Marrch 26. https://www.hirevue.com/blog/hiring/what-is-adverse-impact-and-why-measuring-it-matters. Accessed 18 Sept 2020

24. World Health Organization et al., A. Backurs, P. Indyk, K. Onak, B. Schieber, A. Vakilian, and T. Wagner, in *Scalable Fair Clustering*, ArXiv190203519 Cs, 2019 June, Accessed 18 Sep 2020. [Online]. Available: http://arxiv.org/abs/1902.03519. *Osteoarthr. Cartil*

25. J. Goh, S. Adepu, K.N. Junejo, A. Mathur, A dataset to support research in the design of secure water treatment systems, in *International Conference on Critical Information Infrastructures Security*, (Springer, Cham, 2016), pp. 88–99

26. The battle of the attack detection algorithms (BATADAL) dataset. https://www.batadal.net/. Accessed 18 Sep 2020

27. Cyber Science Lab – Malware Datasets. https://cybersciencelab.org/. Accessed 18 Sept 2020

28. S. Mohammadi, H. Mirvaziri, M. Ghazizadeh-Ahsaee, H. Karimipour, Cyber intrusion detection by combined feature selection algorithm. J. Inf. Secur. Appl. **44**, 80–88 (2019)

29. M. Saharkhizan, A. Azmoodeh, A. Dehghantanha, K.-K.R. Choo, R.M. Parizi, An ensemble of deep recurrent neural networks for detecting IoT cyber attacks using network traffic. IEEE Internet Things J. **7**(9), 8852–8859 (2020). https://doi.org/10.1109/jiot.2020.2996425

30. A. Azmoodeh, A. Dehghantanha, K.-K.R. Choo, Robust malware detection for internet of (battlefield) things devices using deep eigenspace learning. IEEE Trans. Sustain. Comput. **4**(1), 88–95 (2018)

31. A. Al-Abassi, H. Karimipour, A. Dehghantanha, R.M. Parizi, An ensemble deep learning-based cyber-attack detection in industrial control system. IEEE Access **8**, 83965–83973 (2020)

32. A.N. Jahromi, J. Sakhnini, H. Karimpour, A. Dehghantanha, A deep unsupervised representation learning approach for effective cyber-physical attack detection and identification on highly imbalanced data, in *Proceedings of the 29th Annual International Conference on Computer Science and Software Engineering*, (2019), pp. 14–23

33. A. Yazdinejad, R.M. Parizi, A. Dehghantanha, Q. Zhang, K.-K.R. Choo, An energy-efficient SDN controller architecture for IoT networks with blockchain-based security. IEEE Trans. Serv. Comput. **13**, 625 (2020)

34. D. Połap, G. Srivastava, A. Jolfae, R. M. Parizi, Blockchain technology and neural networks for the internet of medical things. Researchers.mq.edu.au; Institute of Electrical and Electronics Engineers (IEEE) (2020). https://doi.org/10.1109/INFOCOMWKSHPS50562.2020.9162735

35. A. Yazdinejad, G. Srivastava, R.M. Parizi, A. Dehghantanha, K.-K.R. Choo, M. Aledhari, Decentralized authentication of distributed patients in hospital networks using Blockchain. IEEE J. Biomed. Heal. Inform. **24**, 2146 (2020)

36. Q. Chen, G. Srivastava, R.M. Parizi, M. Aloqaily, I. Al Ridhawi, An incentive-aware blockchain-based solution for internet of fake media things. Inf. Process. Manag. **57**, 102370 (2020). https://doi.org/10.1016/j.ipm.2020.102370

37. A. Yazdinejad, R.M. Parizi, A. Bohlooli, A. Dehghantanha, K.-K.R. Choo, A high-performance framework for a network programmable packet processor using P4 and FPGA. J. Netw. Comput. Appl. **156**, 102564 (2020)

38. R.M. Parizi, S. Homayoun, A. Yazdinejad, A. Dehghantanha, K.-K.R. Choo, Integrating privacy enhancing techniques into Blockchains using sidechains, in *IEEE Canadian Conference of Electrical and Computer Engineering (CCECE)*, (2019), pp. 1–4. https://doi.org/10.1109/CCECE.2019.8861821

39. A. Yazdinejad, R. M. Parizi, G. Srivastava, A. Dehghantanha, K.-K. R. Choo, Energy efficient decentralized authentication in internet of underwater things using blockchain, in *2019 IEEE Globecom Workshops (GC Wkshps)*, (2019). https://doi.org/10.1109/gcwkshps45667.2019.9024475

40. V. Mothukuri, R.M. Parizi, S. Pouriyeh, Y. Huang, A. Dehghantanha, G. Srivastava, A survey on security and privacy of federated learning. Futur. Gener. Comput. Syst. **115**, 619 (2020)
41. A. Yazdinejad, H. HaddadPajouh, A. Dehghantanha, R.M. Parizi, G. Srivastava, M.-Y. Chen, *Cryptocurrency Malware Hunting: A Deep Recurrent Neural Network Approach*, vol 96 (Elsevier, 2020)
42. M. Aledhari, R. Razzak, R.M. Parizi, F. Saeed, Federated learning: A survey on enabling technologies, protocols, and applications. IEEE Access **8**, 140699–140725 (2020). https://doi.org/10.1109/ACCESS.2020.3013541
43. A. Yazdinejad, A. Bohlooli, K. Jamshidi, Performance improvement and hardware implementation of Open Flow switch using FPGA, in *2019 5th Conference on Knowledge Based Engineering and Innovation (KBEI)* (2019). https://doi.org/10.1109/KBEI.2019.8734914
44. S.M. Tahsien, H. Karimipour, P. Spachos, Machine learning based solutions for security of Internet of Things (IoT): A survey. J. Netw. Comput. Appl. **161**, 102630 (2020)
45. H. HaddadPajouh, A. Dehghantanha, R. Khayami, K.-K.R. Choo, A deep recurrent neural network based approach for internet of things malware threat hunting. Futur. Gener. Comput. Syst. **85**, 88–96 (2018). https://doi.org/10.1016/j.future.2018.03.007
46. H. Darabian et al., Detecting Cryptomining malware: A deep learning approach for static and dynamic analysis. J. Grid Comput., 1–11 (2020)
47. E.M. Dovom, A. Azmoodeh, A. Dehghantanha, D.E. Newton, R.M. Parizi, H. Karimipour, Fuzzy pattern tree for edge malware detection and categorization in IoT. J. Syst. Archit. **97**, 1–7 (2019)
48. A. Azmoodeh, A. Dehghantanha, M. Conti, K.-K.R. Choo, Detecting crypto-ransomware in IoT networks based on energy consumption footprint. J. Ambient. Intell. Humaniz. Comput. **9**(4), 1141–1152 (2018)
49. H.H. Pajouh, R. Javidan, R. Khayami, D. Ali, K.-K.R. Choo, A two-layer dimension reduction and two-tier classification model for anomaly-based intrusion detection in IoT backbone networks. IEEE Trans. Emerg. Top. Comput. **7**, 314 (2016)
50. H. Darabian, A. Dehghantanha, S. Hashemi, S. Homayoun, K.R. Choo, An opcode-based technique for polymorphic internet of things malware detection. Concurr. Comput. Pract. Exp. **32**(6), e5173 (2020)
51. Z. Obermeyer, B. Powers, C. Vogeli, S. Mullainathan, Dissecting racial bias in an algorithm used to manage the health of populations. Science (80-) **366**(6464), 447–453 (2019)
52. J. Sylvester, E. Raff, What About Applied Fairness?, *arXiv Prepr. arXiv1806.05250* (2018)
53. S. Bera, D. Chakrabarty, N. Flores, M. Negahbani, Fair algorithms for clustering, in *Advances in Neural Information Processing Systems*, (2019), pp. 4954–4965. Curran Associates. https://papers.nips.cc/paper/2019/file/fc192b0c0d270dbf41870a63a8c76c2f-Paper.pdf
54. M. Schmidt, C. Schwiegelshohn, C. Sohler, Fair coresets and streaming algorithms for fair k-means clustering, *arXiv Prepr. arXiv1812.10854* (2018)
55. C. Rösner, M. Schmidt, Privacy preserving clustering with constraints, *arXiv Prepr. arXiv1802.02497* (2018)
56. H. Elzayn et al., Fair algorithms for learning in allocation problems, in *Proceedings of the Conference on Fairness, Accountability, and Transparency*, (2019), pp. 170–179
57. A. Dash, A. Shandilya, A. Biswas, K. Ghosh, S. Ghosh, A. Chakraborty, Summarizing user-generated textual content: Motivation and methods for fairness in algorithmic summaries. Proc. ACM Human-Comput. Interact. **3**(CSCW), 1–28 (2019)
58. A. Backurs, P. Indyk, K. Onak, B. Schieber, A. Vakilian, T. Wagner, Scalable fair clustering, *arXiv Prepr. arXiv1902.03519* (2019)
59. k-medoids clustering – MATLAB kmedoids. https://www.mathworks.com/help/stats/kmedoids.html. Accessed 18 Sep 2020

Evaluation of Supervised and Unsupervised Machine Learning Classifiers for Mac OS Malware Detection

Dilip Sahoo and Yash Dhawan

1 Introduction

The number of attacks targeting Mac OS has considerably risen in the past couple of years [1]. It is estimated that reported attacks have exceeded 4 million as of 2018 and another 1.8 million attacks have been reported during the first half of 2019 [2]. The first Mac malware was reported in 2004 with Renepo script worm which disabled Mac OSX security and installed malicious toolkit [3]. Adware and Potential Unwanted Program (PUP) resulted in a serious threat for Mac users over the past couple of years as it resulted in security vulnerability making it more likely to get infected by malware [4, 5]. Though the Mac platform is considered safer than Windows it is still prone to several phishing attacks, java-based exploit, the man in the middle attacks, and should not be considered as a bulletproof operating system [6]. Protecting IT resources and computer hardware [7, 8] against malware threats has become vital for corporations and individuals [9–12]. Most antivirus (AV) software use a signature-based technique to detect the threats. A signature of the known malware like spyware, viruses, trojans, worms is stored in a database and if an attack occurs by them in the future then they can be detected against their stored signatures. However, there are many drawbacks to this approach of malware detection. Firstly, the signature-based approach is ineffective against the new malware that is not known previously. Secondly, the metamorphic malware (a variant of known malware) can bypass the antivirus by changing its signature [13]. Significant improvements have been done to make the AVs more effective using

D. Sahoo
Cyber Science Lab, University of Guelph, Guelph, ON, Canada
e-mail: dilip@cybersciencelab.org

Y. Dhawan (✉)
School of Computer Science, University of Guelph, Guelph, ON, Canada
e-mail: ydhawan@uoguelph.ca

© Springer Nature Switzerland AG 2021
K.-K. R. Choo, A. Dehghantanha (eds.), *Handbook of Big Data Analytics and Forensics*, https://doi.org/10.1007/978-3-030-74753-4_11

more sophisticated analysis techniques in recent years. However, there is still the problem of delay between the detection of new malware and updating the signature databases to counter it. This delay can cause significant damage to corporations [14].

In the last decade, more sophisticated methods are being used by researchers for metamorphic malware detection like dynamic and heuristic analysis. In dynamic analysis, the behavior of the malware program is observed at runtime in an isolated sandbox environment like a virtual box. During this process, specific behaviors of a program like system calls, registry updates, network traffic usage, etc. are monitored and used to classify whether the program is a benign application or malware. However, this method can be time-consuming, and sometimes evasive methods used by the malware can detect the analysis environment and stops the malicious code execution or delay the execution [15]. Another main disadvantage is that dynamic analysis cannot be used in realtime scenarios.

In contrast, the heuristic approach uses Machine learning (ML) to learn the malicious program behaviors and can classify them as malware [16–21]. They are easy to implement and can also effectively detect metamorphic malware. The disadvantage of the heuristic approach is higher false positives i.e. the benign programs incorrectly classified as malware. To overcome this issue, the machine learning classifiers need to be trained with datasets with sufficient features and have a balanced ratio between the majority and minority class. A dataset with a large number of features can decrease the false positives but it can increase the overall computation and processing time and hence it is important to analyze and reduce the dataset dimension. There is the number of automated tools and application available which provide tons of features to neutralize a threat before it can compromise a system [22]. For an effective antimalware solution, many companies have adopted the machine learning approach [23]. ML models are trained to distinguish malicious and benign apps using supervised [14, 24–26] and unsupervised classifiers [19, 27–29]. In the supervised learning method, a labeled dataset of malware and good ware is used for the training of the ML classifier. After sufficient training with labeled data, the ML model is used to classify the unseen samples. Whereas, in the unsupervised learning method, the classification is done on observed similarities or differences [30].

In this paper, we employed a heuristic approach using the same raw Mac OSX dataset used by H. H. Pajouh et al. [31], and a comprehensive study was done to evaluate different machine learning classifiers for the detection of MacOSX based malware samples. Below are the measure research work done as part of the experiment.

I. The experiment applied a (Term Frequency – Inverse Data Frequency) TF-ID based text processing was done to extract 654 new features based on the calling of application libraries.

II. A SMOTE data set was developed with balanced distributions of benign and malware samples to reduce bias in favor of any particular class. Each of the classifiers was evaluated using the SMOTE data and a thorough comparison was made against the original dataset and result.

III. Five different machine learning algorithms (4 Supervised and 1 Unsupervised machine learning technique) namely Logistic regression, Random Forest, Decision Tree, Naïve Bayes, and K-nearest neighbor were evaluated and analyzed from different aspects like accuracy, False Positive rate, processing time.

We used the commonly used matrices for the performance evaluation of the machine learning classifiers used in the experiment i.e. True Positive Rate (TPR), False Positive Rate (FPR), Precision, Recall, F-measure, Receiver Operating Characteristics (ROC), and Area Under the ROC Curve (AUC). Detailed descriptions of the evaluation measures are described in Sects. 4.1 and 4.4.

Section 2 of this paper contains related work from recent years. Section 3 details the methodology used in this work, Sect. 4 discuss the results observed in the experiment, and in Sect. 5 we provide the conclusion of our work and suggest future work. Acknowledgment and References for our work are provided at the very end of the paper.

2 Related Work

The threat landscape for MacOS is changing drastically as the amount of malicious software is growing [2]. The velocity, volume, and complexity of malware are posing numerous challenges for the antivirus companies [1]. Various supervised and unsupervised machine learning techniques are proving to be efficient in the detection of malware. Ransomware attacks rose drastically ever since the introduction of cryptocurrencies through which attackers were able to receive ransom anonymously. The majority of ransomware families have different versions and features which makes their detection and analysis sophisticated [32–38]. To resolve this T. Dargahi [27] research provides the first scientific taxonomy of ransomware features, aligned with the Lockheed Martin Cyber Kill Chain (CKC) model. A comprehensive taxonomy would assist researchers in assessing the vulnerability and attack vectors towards intended victims. Sajad. H [39] proposed DRTHIS: Deep ransomware threat hunting and intelligence system which can detect ransomware activities utilizing Long Short-Term Memory (LSTM) and Convolutional Neural Network (CNN), two deep learning techniques. In the classification of ransomware instances, DRTHIS achieved an F-measure of 99.6% with a true positive rate of 97.2%. DRTHIS accurately predicted the ransomware instances based on sequences of action performed by good ware and ransomware samples to correctly classify ransomware samples.

Fattori et al. [40] developed an AccessMiner behavioral malware protection system that provides a high level of OS protection (around 90% with zero false positives). It generally detects malicious samples in real-time by monitoring interactions between applications and the Windows Operating system. Novel Active Learning (AL) framework introduced by Nissim [41] assisted antivirus vendors in determining more malware samples than the existing AL method. It provided an accuracy of 97% as well as provided an increased efficiency to detect novel Windows malware. To reduce the chance of Malware evasion Mangialardo and Duarte [42] proposed the unification of Static and Dynamic analysis using C5.0 and Random Forests (RF) algorithms with an accuracy of 93% for detecting Linux malware. Saharkhizan et al. [43] utilized deep generative metric learning for identifying complex shape of malware data as model space and applied it on NSL-KDD network attack dataset and could obtain high detection rate for different attacks against network.

Since Windows OS is used extensively, a prominent amount of research work has been conducted as compared to OSX malware detection [44]. H.H Pajouh proposed an OSX code inspection technique using Synthetic Minority Over-sampling Technique (SMOTE) to improve malicious sample size in the dataset which helped to achieve a higher malware detection accuracy of 96% and achieve a lower false alarm rate [31]. Some other relevant research includes researcher Pham Duy Phuc used MacOS a malware analysis framework called Mac-A-Mal to automatically capture malware behavior at user and kernel levels. Mac-A-Mal framework led to the discovery of 71 unknown Adware, 2 keyloggers, and 1 trojan involved in the APT32 OceanLotus. It also provided a Heatmap correlation matrix to analyze the correlation of different malware datasets. The model supported static and dynamic analysis to provide a rich set of Mac malware variants that machine learning classifiers can implement [45]. E. Walkup [46] implemented static executable analysis for the detection of Mac malware using different supervised classification techniques. Information gain was utilized in the dataset to select prominent features to detect OS X malware. Our machine learning classifiers aim to solve the gap that was observed in the above papers.

3   Methodology

This section describes the experiment workflow as shown in Fig. 1. First, we obtained the OSX Malware Detection dataset from the cybersciencelab.org [47] website which had 450 benign samples and 152 malware samples. The raw data was then processed to remove the anomalies. Second, different feature selection and feature extraction techniques were used on the processed data, and in the final stage different machine learning classifiers were used to create a detection model. Each step is described in detail in Sects. 3.1, 3.2, and 3.3 respectively.

Two different experiments were conducted following the above steps. The first experiment was conducted using the actual original data that was used to train and test different ML classifiers. In the second experiment, SMOTE data was used which was developed using the oversampling technique. All the experiments were conducted using Python 2.7.1, Jupyter notebook server version 6.0.1 in a Windows 7 virtual machine with intel i7 (2.20GHz) processor.

Fig. 1 Experiment workflow

Table 1 Variance by dimension

D1	D2	D3	D4	D5	D6	D7	D8	D9	D10	D11	D12	D13
0.345	0.507	0.606	0.682	0.753	0.813	0.862	0.906	0.940	0.966	0.984	0.993	1.000

3.1 Data Preprocessing

In this step, we analyzed the MacOSX Malware Detection raw data and found multiple anomalies were found. Each of these anomalies was removed after a thorough inspection. Some columns were having data in both integers and in hex format, which were converted to decimal format so that the features can be of integer type. The null values and bad data were replaced with the mean value of the respective columns. Later this data was converted to CSV (comma-separated values) format and used in the next phase for the feature selection and feature extraction process.

3.2 Feature Selection and Extraction

After the data preprocessing phase, the most relevant features were selected using two different feature selection techniques. We used *ExtraTreeClassifier* [48] form the scikit-learn library for statistical analysis of the feature importance of the dataset. This analysis gave an idea about which features are more relevant as shown in Fig. 2. We observed that features like LoadDYLIB, bind_size, rebase_size, and ncmds (Descriptions are given in Table 2) have more importance relative to other features to determine the final output and classification.

Principal Component Analysis was used to understand how many dimensions of the data maximize the variance of the dataset. This gave us the idea that the first eight dimensions explain approx. 90% of the variance as shown in Table 1.

Heatmap Correlation Matrix, shown in Fig. 3 was used to understand the correlation between different features in the dataset. We observed that the features like ncmds, sizeofcmds, noloadcmd, rebase_size, and bind_size have a greater correlation with the output compared with other features in the dataset. Segments, SectionsTEXT, and SectionsData have the least correlation with the output variable (Table 2).

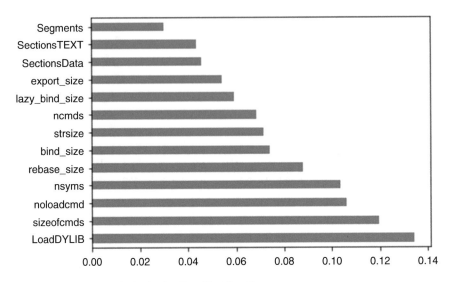

Fig. 2 Feature scores obtained using ExtraTreeClassifier

Table 2 Features description of Mac OS dataset

Feature	Name	Type	Description	Data type
1	ncmds	Integer	No. of commands for every sample Integer	Integer
2	sizeofcmds	Integer	Command size for every sample	Integer
3	noloadcmd	Integer	Number of commands for every loaded sample during execution	Integer
4	rebase_size	Integer	Describe the size of the rebase info Integer	Integer
5	bind_size	Integer	Describing the size of the info to be bound during execution	Integer
6	lazy_bind_size	Integer	States the size of the info to be bound during execution	Integer
7	export_size	Integer	States the size of lazy binding info Integer	Integer
8	nsyms	Integer	States the no of symbol table entries Integer	Integer
9	strsize	Integer	States size of string table in bytes	Integer
10	LoadDYLIB	Integer	States no of DYLIB called and load for execution of malware	Integer
11	Segments	Integer	Number of total segments which consist of every sample	Integer
12	SectionsTEXT	Integer	No of text segments consisting of every sample	Integer
13	SectionsData	Integer	No of data segments consisting of every sample	Integer
14	DYLIBnames	String	Define names of loaded DYLIB	String

Description are adopted from the paper on OSX malware detection

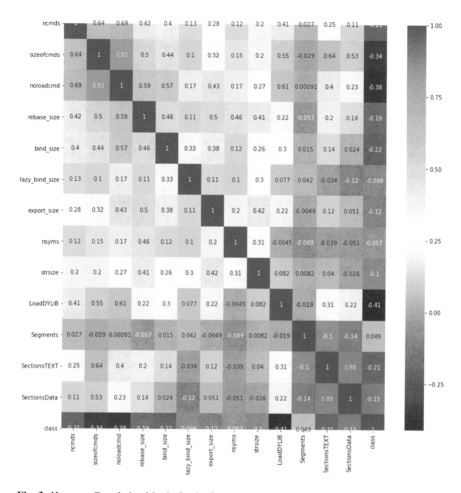

Fig. 3 Heatmap Correlation Matrix for the features

3.2.1 Feature Extraction

The column DYLIBnames in the data set has the information regarding system libraries called by applications. Names of different libraries were stored in a comma-separated string format. We did the text processing of the column separately and extracted several features using *TfidfVectorizer* from scikit-learn (https://scikit-learn.org). During this process, we first used 'comma' as the token separator to split each library's names from the strings in each sample and were then converted to new feature columns. The Term Frequency – Inverse Data Frequency (TFIDF) [49] values of these libraries were calculated and assigned to the new feature columns for each sample. Finally, 654 new features were extracted from the DYLIBnames column and were also used for training and testing the ML classifiers.

Table 3 Original dataset and
the SMOTE dataset sample
distributions

Dataset	Benign	Malicious	Total
Original	460	152	612
SMOTE	460	460	920

3.2.2 SMOTE Dataset Development

Balancing the data is crucial because an imbalanced dataset can cause biased results
in favor of the majority class [50]. Our original dataset contained 450 samples from
the benign class and 152 from malware class which was in the ratio of approx. 3
to 1. Hence, a SMOTE dataset was developed with an oversampling technique to
balance the minority class. Table 3 illustrates the data sample details for original
and SMOTE datasets.

3.3 Machine Learning Classifier Phase

Five different machine learning classifier namely *LogisticRegression, Random Forest, KNN, decision tree,* and *Naïve Bayes* were used where KNN is an unsupervised
classifier, and rest are supervised classifiers. The ML model is trained and tested in
two stages. In the first stage, the original dataset was used along with the 654 newly
extracted features, and in the second stage, a SMOTE enhanced dataset was used to
train and test all the classifiers.

4 Experiment and Results

This section describes the results obtained from our experiments and the performance evaluation of different ML classifiers used during the experiment. The results
were obtained using a tenfold cross-validation technique. The assessments were
done in two phases. First, the actual results from different ML classifiers were
obtained with the original dataset and in the second phase, the SMOTE dataset
was used and results were recorded. Later, more analysis was done by making
comparisons of important performance metrics like Accuracy, ROC curves, False
Positive Rate, and processing time. Finally, the results were compared with another
similar paper.

4.1 Evaluation Measures

Output results of the ML classifiers can be summarised as a confusion matrix where the diagonal elements are True Positives (TP), True Negatives (TN), and the off-diagonal elements as False Positives (FP) and False Negatives (FN) respectively.

True positive (TP): The normal observation is predicted as normal
True Negative (TN): Anomalous observation is predicated as anomalous
False Positive (FP): Anomalous observation is predicted as normal
False Negative (FN): Normal observation is predicated as anomalous

$$\text{True Positive Rate (TPR)} = \frac{TP}{TP + FN}$$

$$\text{False Positive Rate (FPR)} = \frac{FP}{TN + FP}$$

$$\text{Precision} = \frac{TP}{TP + FP}$$

$$\text{Recall} = TPR = \frac{TP}{TP + FN}$$

$$F - \text{measure} = \frac{2 * \text{Recall} * \text{Precision}}{\text{Recall} + \text{Precision}}$$

$$\text{Accuracy} = \frac{TP + TN}{TP + TN + FP + FN}$$

Here TPR, also known as Recall is the value of predicted malware classified correctly and FPR is the value of normal data incorrectly predicted as malware. Precision is also known as the positive predicted value, returns the rate of relevant results. F-measure provides value that estimates the entire system performance by combining precision and recall into a single number. Accuracy denotes how accurately an ML classifier can classify the binary classes i.e. 'good ware' and 'malware'.

4.2 Evaluation of ML Classifiers

We have used different performance metrics namely: Precision, Recall, F-measure, and Accuracy to evaluate our ML models. Table 4 shows a summary of the overall results obtained from the two experiments i.e. using normal dataset and the SMOTE dataset. In the first experiment, Logistic regression achieved the highest detection rate of 0.94 in terms of testing accuracy. Random Forest and then KNN came next and achieved an accuracy of 0.93 and 0.88 respectively.

In the second experiment, an overall increment in all parameters of the detection result was observed for the five classifiers when the SMOTE dataset was used. Especially, for Decision Tree and Naïve Bayes, a significant increment of approx. 10% was observed in the accuracy. In the second experiment also Logistic Regression remained on the top with a detection accuracy of 0.96 followed by Random Forest with 0.95, Decision Tree with 0.94, Naïve Bayes with 0.93, and KNN with 0.90. Surprisingly, Decision Tree and Naïve Bayes outperformed KNN in terms of all the parameters i.e. precision, Recall, F-measure, and testing accuracy during this experiment.

4.3 False-Positive Rate Comparison

Table 5 shows the count of predicted TP, FN, FP, TN while testing each of the ML classifiers. Then, FPR was calculated and compared for both the experiments using original data and SMOTE data respectively. A significant reduction in FPR was recorded for the SMOTE dataset compared to the original dataset.

Figure 4 shows the false-positive rate (FPR) comparison between the two experiments using the above mentioned different datasets for each classifier. It was observed that with the original dataset Random Forest had the highest false-positive

Table 4 Result summary comparing evaluation result from the Original dataset and the SMOTE dataset

Experiment	No.	Classifier	Precision	Recall	F-measure	Accuracy
I. Original dataset	1	Logistic Regression	0.94	0.94	0.94	0.94
	2	Random Forest	0.94	0.93	0.93	0.93
	3	KNN	0.89	0.88	0.88	0.88
	4	Decision Tree	0.87	0.87	0.87	0.87
	5	Naïve Bayes	0.86	0.83	0.84	0.83
II. SMOTE dataset	1	Logistic Regression	0.96	0.96	0.96	0.96
	2	Random Forest	0.95	0.95	0.95	0.95
	3	KNN	0.91	0.9	0.89	0.90
	4	Decision Tree	0.94	0.94	0.93	0.94
	5	Naïve Bayes	0.94	0.93	0.93	0.93

Table 5 TP, TN, FP, TN count, and FPR value for ML classifiers tested with the original dataset and SMOTE dataset

Experiment	No.	Classifier	TP	FN	FP	TN	FPR
I. Original data	1	Logistic Regression	91	3	4	25	0.14
	2	Random Forest	94	0	8	21	0.28
	3	KNN	83	11	4	25	0.14
	4	Decision Tree	85	9	7	22	0.24
	5	Naïve Bayes	78	16	5	24	0.17
II. SMOTE data	1	Logistic Regression	129	8	3	137	0.02
	2	Random Forest	128	9	5	135	0.04
	3	KNN	109	28	1	139	0.01
	4	Decision Tree	126	11	7	133	0.05
	5	Naïve Bayes	117	20	0	140	0.00

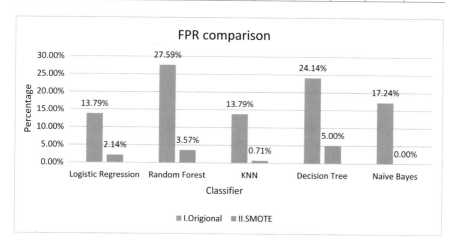

Fig. 4 False alarm (FPR) comparison of the ML classifiers for Original dataset and SMOTE dataset

rate i.e. 27.59%, followed by Decision Tree at 24.14%. For both KNN and Logistic Regression, the FPR was relatively low at 13.79%. After running the ML classifiers on the SMOTE dataset, there was a significant decrease in the FPR value for all the classifiers. Naive Bayes achieved a noteworthy 0% FPR rate with SMOTE dataset. The difference in false alarm for a small amount of data won't be significant, but practically the traffic over any common network is huge enough that the difference can be observed.

4.4 ROC Curve

Figures 5 and 6 shows the comparison of the Receiver Operating Characteristic (ROC) Curve for the ML Classifiers. ROC curve [51] is one of the methods of measuring the performance of a classification model. In this curve, the True Positive Rate (TPR) is plotted against False Positive Rate (FPR) for the probabilities of the classifier predictions. Then, the area under the plot is calculated. More the area under the curve, better is the model at distinguishing between classes.

In Figs. 5 and 6, the graphs illustrate that the Logistic Regression has the highest AUC of 0.987 and 0.993 for the original dataset and SMOTE dataset respectively, which signifies an outstanding prediction score. AUC for RF is was recorded a little lower than Logistic Regression at 0.971 and 0.989 for the mentioned datasets denoting very good prediction as well. Finally, followed by KNN, Decision Tree, and Naïve Bayes for the same datasets. It was observed that there was an increment in AUC value for each of the ML classifiers when the SMOTE dataset was used.

4.5 Performance Evaluation

In this section performance evaluation of each classifier in terms of processing time is discussed. Figure 7 shows the execution time comparison between the two experiments using the original data set and SMOTE dataset. First, with the original dataset, Logistic Regression and Naïve Bayes performed better than the rest of the

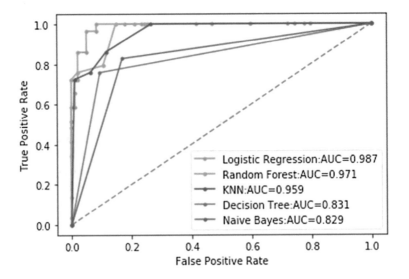

Fig. 5 ROC curve for the ML classifiers for Original dataset

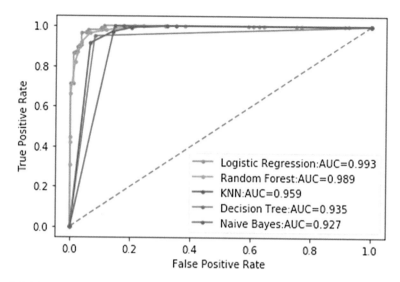

Fig. 6 ROC curve for the ML classifiers for SMOTE dataset

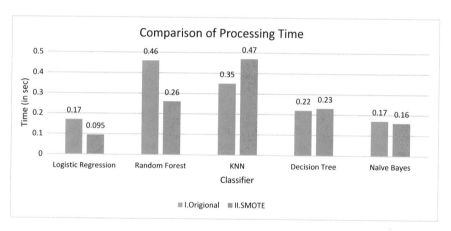

Fig. 7 performance comparison based on the execution time for the ML classifiers

algorithms, where both took 170 ms. In experiment-I, Random Forest took the highest execution time of 460 ms. Later, with the SMOTE dataset again Logistic Regression and Naïve Bayes performed better than other classifiers and a slight decrease in execution time was also noted for them. For KNN an increase in the execution time was recorded when SMOTE dataset was used, where it took 470 ms compared to 350 ms in the first experiment. Other than KNN the execution time either reduced or remained almost the same in the second experiment with the SMOTE dataset.

4.6 *Result Comparison*

Authors in [31] researched macOS malware detection using the same dataset. They used three different SMOTE enhanced i.e. 2x_SMOTE, 3x_SMOTE, and 5x_SMOTE dataset for their experiment. However, the benign and malware sample was always in the ratio of approx. 3:1 in the datasets like it was in the original dataset. The sample weight of libraries was used with an occurrence probability value for the new feature creation. Decision achieved the highest accuracy of 96% and 4% false alarm with the 5x_SMOTE dataset. Weighted RBF-SVM achieved 91% accuracy with a 3.9% false alarm using the original dataset.

In this paper, we used SMOTE data with a balanced distribution for both benign and malware samples in the ratio of 1:1. The library calls frequency was calculated using TF-IDF text processing method and was used for the new feature creation. In our experiment, the highest accuracy was obtained from Logistic regression which was 96% with a false alarm of 2.14% using SMOTE data. Using original data, Logistic regression achieved the highest accuracy of 94% and a false alarm of 13.79%. Overall, all the classifiers used in this paper show better detection accuracy and a lower false-positive rate.

5 Conclusion and Future Work

In this paper, we used TF-ID based text processing to extract new features, and using the Synthetic Minority Over-sampling Technique (SMOTE) we were able to balance our dataset and achieve better results compared to our original dataset. We tested our dataset with 5 machine learning classifiers (4 Supervised and 1 unsupervised algorithm) and achieved a promising accuracy of 96% for Logistic regression and 95% for Random forest. The experiment also provided us with a lower false-positive rate as evident in Fig. 7.

Future work would include malware detection and analysis to be further improved by acquiring more data samples to create a better detection system. Additional research can be done on a larger dataset to predict if the same results are obtained or not. To acquire predictive performance ensemble machine learning technique works well-using bagging and boosting techniques using several base models. With Ensemble machine learning we can reduce variance, noise, and bias as well as increases the accuracy of the model [52]. A deep learning approach can also be used on a large malware dataset to detect complex malware and outperform traditional machine learning algorithms [53]. This paper can be treated as a case study for researchers working in the area of intelligent MacOS malware detection systems on enterprise and cloud platforms.

References

1. J. Stoldt, T. Uwe Trapp, Toussai, Mac malware getting serious – Security no longer optional (Macworld, 2018, Februrary 5), https://www.macworld.com/article/3253252/mac-malware-getting-serious-security-no-longeroptional.html. Accessed 24 Oct 2019
2. Threats to macOS users, https://securelist.com/threats-to-macos-users/93116/. Accessed 22 Oct 2019
3. Mac malware facts, https://www.eset.com/int/mac-malware-facts/. Accessed 10 May 2020
4. World Health Organization, et al., T. Reed, Mac security facts and fallacies (Malwarebytes Labs, 2017, March 8). https://blog.malwarebytes.com/101/2017/03/mac-security-facts-and-fallacies/. Accessed 10 May 2020. *Osteoarthr. Cartil*
5. A. Yazdinejad, R.M. Parizi, A. Dehghantanha, K.-K.R. Choo, P4-to-blockchain: A secure blockchain-enabled packet parser for software defined networking. Comput. Secur. **88** (2020). https://doi.org/10.1016/j.cose.2019.101629
6. Warning as Mac malware exploits climb 270% | Computerworld, https://www.computerworld.com/article/3262225/warning-as-mac-malware-exploits-climb-270.html. Accessed 10 May 2020
7. A. Yazdinejad, A. Bohlooli, K. Jamshidi, Efficient design and hardware implementation of the OpenFlow v1.3 Switch on the Virtex-6 FPGA ML605. J. Supercomput. **74**(3) (2018). https://doi.org/10.1007/s11227-017-2175-7
8. A. Yazdinejad, A. Bohlooli, K. Jamshidi, Performance improvement and hardware implementation of Open Flow switch using FPGA, in *IEEE 5th Conference on Knowledge Based Engineering and Innovation, KBEI 2019*, (2019), pp. 515–520
9. A. Dehghantanha, K.-K. R. Choo (eds.), *Handbook of Big Data and IoT Security* (Springer, Cham, 2019)
10. A. Azmoodeh, A. Dehghantanha, Big data and privacy: Challenges and opportunities, in *Handbook of Big Data Privacy*, (Springer, Cham, 2020), pp. 1–5. https://doi.org/10.1007/978-3-030-38557-6_1
11. M. Conti, A. Dehghantanha, K. Franke, S. Watson, Internet of Things security and forensics: Challenges and opportunities. Futur. Gener. Comput. Syst. **78**, 544–546 (2018). https://doi.org/10.1016/j.future.2017.07.060
12. M. Conti, T. Dargahi, A. Dehghantanha, Cyber threat intelligence: Challenges and opportunities, in *Advances in Information Security*, (Springer, Cham, 2018), pp. 1–6. https://doi.org/10.1007/978-3-319-73951-9_1
13. Y. Ye, T. Li, Q. Jiang, Y. Wang, CIMDS: Adapting postprocessing techniques of associative classification for malware detection. IEEE Trans. Syst. Man Cybern. Part C Appl. Rev. **40**(3), 298–307 (2010)
14. S. Homayoun, A. Dehghantanha, M. Ahmadzadeh, S. Hashemi, R. Khayami, Know abnormal, find evil: Frequent pattern mining for ransomware threat hunting and intelligence. IEEE Trans. Emerg. Top. Comput. (2017). https://doi.org/10.1109/TETC.2017.2756908
15. S.K. Sahay, A. Sharma, Grouping the executables to detect malware with high accuracy. arXiv Prepr. arXiv1606.06908 (2016)
16. A. Azmoodeh, A. Dehghantanha, M. Conti, K.-K.R. Choo, Detecting crypto-ransomware in IoT networks based on energy consumption footprint. J. Ambient. Intell. Humaniz. Comput. **9**(4), 1141–1152 (2018)
17. A. Azmoodeh, A. Dehghantanha, K.-K.R. Choo, Robust malware detection for internet of (battlefield) things devices using deep eigenspace learning. IEEE Trans. Sustain. Comput. **4**(1), 88–95 (2018)
18. H. Haddadpajouh, A. Azmoodeh, A. Dehghantanha, R.M. Parizi, MVFCC: A multi-view fuzzy consensus clustering model for malware threat attribution. IEEE Access **8**, 139188–139198 (2020)
19. H.H. Pajouh, R. Javidan, R. Khayami, D. Ali, K.-K.R. Choo, A two-layer dimension reduction and two-tier classification model for anomaly-based intrusion detection in IoT backbone networks. IEEE Trans. Emerg. Top. Comput. **7**(2), 314–323 (2016)

20. M. Saharkhizan, A. Azmoodeh, A. Dehghantanha, K.-K.R. Choo, R.M. Parizi, An ensemble of deep recurrent neural networks for detecting IoT cyber attacks using network traffic. IEEE Internet Things J. **7**(9), 8852–8859 (2020). https://doi.org/10.1109/jiot.2020.2996425

21. A.N. Jahromi et al., An improved two-hidden-layer extreme learning machine for malware hunting. Comput. Secur. **89**, 101655 (2020)

22. P.N. Bahrami, A. Dehghantanha, T. Dargahi, R.M. Parizi, K.-K.R. Choo, H.H.S. Javadi, Cyber kill chain-based taxonomy of advanced persistent threat actors: Analogy of tactics, techniques, and procedures. J. Inf. Process. Syst. **15**(4), 865–889 (2019)

23. How machine learning works | Kaspersky official blog, https://www.kaspersky.com/blog/machine-learning-explained/13487/. Accessed 10 May 2020

24. A. Azmoodeh, A. Dehghantanha, R.M. Parizi, H. Karimipour, E. Modiri, D.E. Newton, Fuzzy pattern tree for edge malware detection and categorization in IoT zero trust distributed computing view project naive-Bayesian-based model for interoperability among heterogeneous Systems in Intelligent Buildings View project fuzzy pattern tree for. Art. J. Syst. Arch. **97**, 1–7 (2019)

25. H. HaddadPajouh, A. Dehghantanha, R. Khayami, K.-K.R. Choo, A deep recurrent neural network based approach for Internet of Things malware threat hunting. Futur. Gener. Comput. Syst. **85**, 88–96 (2018). https://doi.org/10.1016/j.future.2018.03.007

26. H. Darabian et al., A multiview learning method for malware threat hunting: Windows, IoT and android as case studies. World Wide Web **23**(2), 1241–1260 (2020)

27. T. Dargahi, A. Dehghantanha, P.N. Bahrami, M. Conti, G. Bianchi, L. Benedetto, A cyber-kill-chain based taxonomy of crypto-ransomware features. J. Comput. Virol. Hacking Tech. **15**(4), 277–305 (2019)

28. A. Azmoodeh, A. Dehghantanha, R.M. Parizi, S. Hashemi, B. Gharabaghi, G. Srivastava, Active spectral botnet detection based on eigenvalue weighting, in *Handbook of Big Data Privacy*, (Springer, Cham, 2020), pp. 385–397. https://doi.org/10.1007/978-3-030-38557-6_19

29. M. Alaeiyan, A. Dehghantanha, T. Dargahi, M. Conti, S. Parsa, A multilabel fuzzy relevance clustering system for malware attack attribution in the edge layer of cyber-physical networks. ACM Trans. Cyber-Phys. Syst. **4**(3), 1–22 (2020)

30. World Health Organization, et al., E. McNulty, What's the difference between supervised and unsupervised learning? (Dataconomy, 2015, January 8), https://dataconomy.com/2015/01/whats-the-difference-betweensupervised-and-unsupervised-learning/. Accessed 10 May 2020. *Osteoarthr. Cartil*

31. H.H. Pajouh, A. Dehghantanha, R. Khayami, K.-K.R. Choo, Intelligent OS X malware threat detection with code inspection. J. Comput. Virol. Hacking Tech. **14**(3), 213–223 (2018)

32. A. Yazdinejad, R.M. Parizi, G. Srivastava, A. Dehghantanha, K.-K.R. Choo, Energy efficient decentralized authentication in internet of underwater things using blockchain, in *2019 IEEE Globecom Workshops (GC Wkshps)*, (2019), pp. 1–6

33. A. Yazdinejad, H. HaddadPajouh, A. Dehghantanha, R.M. Parizi, G. Srivastava, M.-Y. Chen, Cryptocurrency malware hunting: A deep recurrent neural network approach. Appl. Soft Comput. Elsevier **96**, 106630 (2020)

34. M. Aledhari, R. Razzak, R.M. Parizi, F. Saeed, Federated learning: A survey on enabling technologies, protocols, and applications. IEEE Access **8**, 140699–140725 (2020). https://doi.org/10.1109/ACCESS.2020.3013541

35. A. Yazdinejad, R.M. Parizi, A. Dehghantanha, H. Karimipour, G. Srivastava, M. Aledhari, Enabling drones in the internet of things with decentralized blockchain-based security. IEEE Internet Things J., 1 (2020). https://doi.org/10.1109/jiot.2020.3015382

36. V. Mothukuri, R.M. Parizi, S. Pouriyeh, Y. Huang, A. Dehghantanha, G. Srivastava, A survey on security and privacy of federated learning. Futur. Gener. Comput. Syst. **115**, 619–640 (2020)

37. R.M. Parizi, S. Homayoun, A. Yazdinejad, A. Dehghantanha, K.-K.R. Choo, Integrating privacy enhancing techniques into blockchains using sidechains, in *The Annual IEEE Canadian Conference on Electrical and Computer Engineering*, (2019). https://doi.org/10.1109/CCECE.2019.8861821

38. A. Yazdinejad, R.M. Parizi, A. Dehghantanha, G. Srivastava, S. Mohan, A.M. Rababah, Cost optimization of secure routing with untrusted devices in software defined networking. J. Parallel Distrib. Comput. (2020). https://doi.org/10.1016/j.jpdc.2020.03.021
39. S. Homayoun et al., DRTHIS: Deep ransomware threat hunting and intelligence system at the fog layer. Futur. Gener. Comput. Syst. **90**, 94–104 (2019). https://doi.org/10.1016/j.future.2018.07.045
40. A. Fattori, A. Lanzi, D. Balzarotti, E. Kirda, Hypervisor-based malware protection with accessminer. Comput. Secur. **52**, 33–50 (2015)
41. N. Nissim, R. Moskovitch, L. Rokach, Y. Elovici, Novel active learning methods for enhanced PC malware detection in windows OS. Expert Syst. Appl. **41**(13), 5843–5857 (2014)
42. R.J. Mangialardo, J.C. Duarte, Integrating static and dynamic malware analysis using machine learning. IEEE Lat. Am. Trans. **13**(9), 3080–3087 (2015)
43. M. Saharkhizan, A. Azmoodeh, H. HaddadPajouh, A. Dehghantanha, R.M. Parizi, G. Srivastava, A hybrid deep generative local metric learning method for intrusion detection, in *Handbook of Big Data Privacy*, (Springer, Cham, 2020), pp. 343–357. https://doi.org/10.1007/978-3-030-38557-6_16
44. Operating system guide: Windows vs Mac (vs Linux), https://www.logicalincrements.com/articles/build-pc-windows-apple-mac-linux-operating-system-os. Accessed 10 May 2020
45. D.-P. Pham, D.-L. Vu, F. Massacci, Mac-A-Mal: macOS malware analysis framework resistant to anti evasion techniques. J. Comput. Virol. Hacking Tech. **15**(4), 249–257 (2019)
46. E. Walkup, Mac malware detection via static file structure analysis. Univ. Stanf. [Online]. Available: http://cs229.stanford.edu/proj2014/Elizabeth%20Walkup%20MacMalware.pdf
47. CSL-Home – Cyber Science Lab, https://cybersciencelab.org/. Accessed 10 May 2020
48. 3.2.4.3.3. sklearn.ensemble.ExtraTreesClassifier – Scikit-learn 0.22.2 documentation, https://scikit-learn.org/stable/modules/generated/sklearn.ensemble.ExtraTreesClassifier.html. Accessed 10 May 2020
49. How to process textual data using TF-IDF in Python (freeCodeCamp.org, 2018, June 6), https://www.freecodecamp.org/news/how-to-process-textual-data-using-tf-idf-in-pythoncd2bbc0a94a3/. Accessed 10 May 2020
50. K. Mahendru, How to deal with imbalanced data using SMOTE (Medium, 2019, June 26), https://medium.com/analytics-vidhya/balance-your-data-using-smote-98e4d79fcddb. Accessed 10 May 2020
51. Machine learning classifier evaluation using ROC and CAP curves, https://towardsdatascience.com/machine-learning-classifier-evaluation-using-roc-and-cap-curves-7db60fe6b716. Accessed 10 May 2020
52. R.R.F. DeFilippi, Boosting, bagging, and stacking – Ensemble methods with sklearn and mlens (Medium, 2018, August 4), https://medium.com/@rrfd/boosting-bagging-and-stacking-ensemblemethods-with-sklearn-and-mlens-a455c0c982de. Accessed 10 May 2020
53. R. Vinayakumar, M. Alazab, K.P. Soman, P. Poornachandran, S. Venkatraman, Robust intelligent malware detection using deep learning. IEEE Access **7**, 46717–46738 (2019)

Evaluation of Machine Learning Algorithms on Internet of Things (IoT) Malware Opcodes

Adesola Anidu and Zibekieni Obuzor

1 Introduction

With recent advancements in technology, various mobile devices, gadgets and, applications have been designed and developed to make life easier. The power of the internet has been harnessed in the operation of these devices to make them smarter than ever before. Devices, gadgets connected to the internet in such a way that information can be sent and received from the devices are popularly referred to as the Internet of Things (IoT) devices [1–6]. These devices are being used in various domains such as healthcare, agriculture, military, etc. [7–14]. With various sensors deployed on these devices. Due to the increasing versatility and popularity of these devices, they are prone to malware attacks as most of these devices lack security protection [6, 15–18]. Kaspersky Lab. in 2016 observed that most IoT devices tested, had unpatched vulnerabilities or default passwords making them insecure and hence prone to malicious attacks [19]. These attacks can be eavesdropping, spoofing attack, jamming, Denial of Service (DoS), intrusion, and malicious software (malware) [20]. The most common of these attacks is the malware [21].

Two main methods which are commonly used for the analysis of malware are dynamic and static analysis [22]. In dynamic analysis, the program is run in an emulator or instrumented hardware with to extract characteristic actions executed by the program [23]. It is slower and less prone to code obfuscation. In static analysis, the program binary is disassembled to extract the features [24]. There is greater code coverage in static analysis than dynamic analysis. Some research works have been done to combine the two methods [25]. Analysis of malware aids

A. Anidu (✉) · Z. Obuzor
School of Computer Science, University of Guelph, Guelph, ON, Canada
e-mail: aanidu@uoguelph.ca; zobuzor@uoguelph.ca

© Springer Nature Switzerland AG 2021
K.-K. R. Choo, A. Dehghantanha (eds.), *Handbook of Big Data Analytics and Forensics*, https://doi.org/10.1007/978-3-030-74753-4_12

in understanding the behavior and structure of the malware. Many features can be extracted from the malware executables to identify malware [26]. These features include byte n-gram feature, opcode, function-based, string-based, API calls, and system calls [27, 28]. Byte n-gram features are sequences of n-byte extracted from malware as a signature for recognizing malware [29]. Malware detection using this method has yielded a high accuracy as presented by [25]. String features are features that are based on plain text encoded in executables such as windows, library, etc. String features are not so robust as they can be easily modified though accuracy is better than byte n-gram and portable executables [25]. Portable executables are features extracted from some parts of the exe file. These features show that a file has been manipulated or infected to perform malicious activity [30]. Opcode is an assembly language instruction used to describe the type of operation being performed. Opcode is a mnemonic for operational code [31]. Opcodes can be used to derive variability between malware and a benign ware. Studies have shown that the opcodes feature extraction method is more efficient in classifying malware. According to [25] opcodes reveal a lot of statistical differences between legitimate software and malware. The use of opcodes has been employed in detecting and classifying malware in android mobile devices, but little research work has been done in its use for the detection and classification of IoT malware. A recent study by [19] in the use of opcodes for IoT malware threat hunting using deep neural network yielded an accuracy of 98%. In this study, we will be implementing several machine learning classifiers to achieve higher accuracy in the classification of malware based on the opcodes.

The next section briefly describes related works on the topic. Section 3 describes the methodology of the study. In Sect. 4, the evaluation and experimental results are presented. Lastly, Sect. 5 concludes the paper and presents the future research direction.

2 Literature Review

Various methods such as machine learning approaches [32] have been employed for malware classification and detection in devices. [33] presented a paper to explore the possibility of using a random forest classifier for the detection of malware in an Android device by examining the behavior of the application data. The dataset used was generated automatically by monitoring some attributes on the Android adb-monkey. Random Forest classifier provided an accuracy with fivefold cross-validation of over 99.9% of the correctly classified samples. 0.0171 was obtained as the optimal square root of the mean-squared-error. 0.002 was achieved as the optimal out-of-bag error rate with 40 trees as minimum forest size. Two machine learning aided approaches are presented by [27] for static analysis of mobile applications. One is based on source code analysis using a bag of words representation model while the other is based on permissions. Source code-based method achieved an F-measure of 95.1% while the permission-based approach achieved an F-measure of 89%.

MODroid, a novel Android behavioral-based malware detection technique presented by [34] which comprises of a lightweight client agent and server analyzer. A signature for each application based on the system request of the application is generated by the server analyzer. The signature generated is normalized to improve accuracy. Identification of malware with similarity to behavior signatures of the already generated blacklist of malware signatures is done using Spearman's rank correlation coefficient. The detection rate of 60.16% with 0.4% false-negatives and 39.43% false positives were achieved at a threshold value of 0.90 when experiments were run using MODroid on Genome dataset and APK submissions of Android client agent. [35] presented an Android malware detection model based on permission. This model has two layers. An improved random forest algorithm was used to analyze in the first layer detection while sensitive permission rules that match the fuzzy sets generated in the first layer detection were used for second layer detection. The results of experiments showed that the accuracy rate is not high due to static detection. A hybrid model based on Convolutional Neural Network (CNN) and Deep Autoencoder (DAE) is proposed by [36]. Reconstruction of the high dimensional features of the Android application is done to improve the accuracy in the detection of malware and then CNN is used for the detection of Android malware. DAE is used for the pre-training of CNN to reduce the training time. Experiments were conducted using 10,000 benign apps and 13,000 malicious apps. Results from experiments showed that the training time using DAE-CNN model was reduced by 83% when compared with the serial convolutional neural network.

MalDozer an automatic framework for android malware detection using deep learning is presented by [37]. By using the raw sequence of the app's API method calls, MalDozer extracts and learns the benign and malicious patterns to detect Android malware. F1-score of 96%–99% and a false positive rate of 0.06%–2% were achieved when MalDozer was evaluated using multiple Android malware datasets ranging from 1 k to 33 K malware apps, and 38 K benign apps. An Android malware characterization and detection approach using weight-adjusted deep learning is presented by [38]. An accuracy of over 90% is obtained when evaluated with only 237 features. Two end-to-end malware detection for android IoT devices using deep learning is proposed by [39]. The inputs used in the proposed method are resamples of the raw bytecodes of the classes.dex files of Android applications. An accuracy of 93.4% and 95.8% was achieved using a dataset with 8 K benign applications and 8 k malicious applications. The proposed method has low resource consumption, does not need manual feature engineering, is not limited by input size, and is suitable for Android IoT devices. [40] presents a novel framework for the detection of android malware. A multimodal deep learning method is proposed as a model for malware detection. The performance of the model was evaluated using 41,260 samples which confirmed that the framework was effective for the detection of Android malware. [41] used words from an apk is used to generate the image, which is then analyzed using Convolutional Neural

Network to classify whether the apk file is a malware or not. An accuracy of 92% was achieved.

Malware analysis and classification using Support Vector Machine is presented by [42]. The experiment was conducted on the heterogeneous malware data retrieved from N6 platform and a classification accuracy between 94% to 95% was obtained. Analysis of Malware behavior based on type classification using machine learning is presented by [43]. Here a system is developed as a pre-filtering application in which known malware are filtered from new malware thus increasing detection. Experiments were conducted on 42,068 samples using a random forest algorithm with 160 trees and satisfactory results were obtained. [44] proposed the use of Hidden Markov Model for training using dynamic features to perform malware classification. The features used are observation sequences that are made up of system call traces. Experiments were performed on a dataset containing behavioral profiles of 964 malware programs belonging to 7 different malware families and 50 benign programs. An accuracy of 97% was achieved. [45] presented the use of Artificial Neural Network for analyzing and classifying malware. In the proposed method, the malware is presented as a 2-Dimensional grayscale image. These images as well as their texture similarity were collected for all available variants. Gabor Wavelet transform and GIST were employed to identify the behavior of malicious data using global features. The experiment was conducted using Mahenhur dataset that consists of 3131 binary samples comprising 24 unique families. An accuracy of 96.35% was obtained in detecting and classifying malware using feed-forward Artificial Neural Network.

Furthermore, a deep learning approach for the detection of crypto-mining malware using dynamic and static analysis is presented by [23]. System call events of 1500 Portable Executable (PE) were captured for dynamic analysis. An accuracy of 95% and 99% were achieved for static and dynamic analysis respectively using Long Short-Term Memory (LSTM), Attention-based LSTM (ATT-LSTM), and convolutional neural networks (CNN) approaches on sequences of system call events. [24] proposed a supervised machine learning model for malware threat detection in Mac OS X. Kernel base Support Vector Machine in addition to a novel, weighting measure based on application calls is used to detect the OS X malware. A dataset consisting of 450 benign and 152 malware samples was used to evaluate the model. Results obtained showed that the Synthetic Minority Over-sampling Technique (SMOTE) performed better than the common supervised machine learning algorithm with an accuracy of 96% and a false alarm of less than 4%. An approach for dynamic malware detection based on a combination of Process Mining and Fuzzy Logic techniques is presented by [46]. When the two techniques are combined, a fingerprint of an application is obtained to verify whether it belongs to a known malware family and subsequently identify the difference in detected malware behavior and other variants of malware. Results from experiments conducted on the dataset of 3000 trusted and malicious applications across 12 malware families show good performance in the detection of malware and family identification.

Operational Codes (OpCodes) from programs have been introduced as a reliable feature for identifying and detecting malware using machine learning in devices. A new malware detection method based on OpCodes in an executable file is presented by [47]. A graph of OpCode within an executable file is generated. This graph is subsequently embedded into the eigenspace using the Power Iteration method. Executable files are presented as a linear combination of eigenvectors proportionate to their respective eigenvalues. The proposed method when evaluated using SVM and KNN has a high detection rate, low false-positive rate, and acceptable computational complexity. Malware detection using a control flow-based opcode behavior was presented by [48]. Executable opcode behaviors were extracted using a control flow-based method. A control flow graph for the program is created to determine the opcode behavior of the program from the execution paths when the graph is traversed. Results presented showed a higher accuracy rate and a lower false-positive rate when compared with other text-based detection methods. [49] used the opcode sequence in detecting malicious Android applications. Binary occurrences of k-grams were used instead of the frequency of opcode sequence. Minimal functionalities required for a program to function were used instead of the total number of opcode-sequences. The experiment was conducted using the Genome Project dataset with an accuracy of 96.83% of detecting a malicious application.

OpCodes have been combined with header information, ByteCodes, API calls, attacker's intent, and permissions to create a multi-view learning method for hunting malicious programs [26].

Weights are assigned automatically to different views to increase detection in various environments. This method is the first malware threat-hunting method that can be implemented across different platforms. Results from experiments conducted across IoT, Windows, and Android platforms show high accuracy with a low rate of false positives. [50] presented N-opcode analysis for Android malware classification and categorization. The dex files were disassembled using baksmali. During the feature selection stage, the information gain of each feature was measured. Four machine learning algorithms namely: Naïve Bayes, Random Forest, Partial Decision Tree, and Support Vector Machine were employed using WEKA as the framework. The Android Malware Genome project dataset was used. Experiments conducted on 2520 samples using up to 10-g opcode features indicated that an F-measure of 98% can be achieved.

Furthermore, [51] used Support Vector Machine for OpCode density-based detection of crypto-ransomware. 443 opcodes were extracted from samples using static analysis of benign and malicious Portable Executable files. These opcodes were represented as density histograms in the dataset. A precision of 100% was achieved in differentiating goodware from ransomware. [52] classified ransomware families with machine learning using the N-gram of opcodes. Opcode sequences generated from the ransomware samples were converted to N-gram sequences. For each N-gram sequence, Term Frequency-Inverse Document Frequency (TF-IDF) is calculated to choose the feature N-grams which exhibit better discrimination between families. Five machine learning algorithms were employed on the feature vectors generated from the feature N-grams for ransomware classification. Results obtained showed an accuracy of 91.43% in classifying ransomware with the F1-measure of the ransomware family up to 99%. The accuracy of the binary classification is 99.3%. [53] proposed an approach for detecting mobile malware using an opcode frequency histogram. Malware is classified using a set of features that count the frequency of a specific group of opcodes extracted from the smali dalvik code of the particular application under analysis. The input data is presented in the form of a histogram representing the opcodes (6 of them) within each class. The distance between the histograms was computed. Classification analysis was conducted using Weka. Experiments were conducted on 5560 Android trusted application and malware applications collected from the Drebin project using six classification algorithms namely: J48, Random Tree, LadTree, Random Forest, NBTree, and RepTree. The results obtained showed that the proposed method produced over 93% precision rate in the detection of mobile malware with an accuracy of 95%.

Also, [24] used a sequential pattern mining technique to detect the most frequent opcode sequences of malicious IoT applications. Sequential pattern mining algorithms were used in the extraction of sub-sequences embedded in the text given which is based upon a support value and a user-specified threshold. Sequential pattern mining was combined with machine learning techniques to classify IoT goodware, malware, and polymorphic malware samples. The dataset used was made up of 269 IoT good ware and 247 malware samples. 36 features were detected because of frequent patterns of malware opcodes and the division to transitional and atomic types. An accuracy and F-measure of above 99% were achieved in detecting IoT malware from benign samples using k-nearest neighbor, MLP, SVM, random forest, Adaboost, and Decision Tree machine learning classifiers. [30] presented a modified Two-hidden-layered Extreme Learning Machine (TELM) for malware hunting. TELM utilizes the dependency of malware sequence elements as well as avoidance of backpropagation when training neural networks. In comparison with the stacked Long Short Term Memory (LSTM) and Convolutional Neural Network (CNN), the proposed method accelerates the training and detection steps for malware hunting. An accuracy of 99.65% was obtained in detecting samples of IoT malware when the proposed approach was evaluated using an IoT-specific dataset. The proposed approach can be utilized on all platforms for malware analysis.

A lightweight classification of IoT malware using Image Recognition is proposed by [54]. One-channel gray-scale images that were converted from binaries were extracted. A light-weight convolutional neural network is implemented to classify IoT malware families. Experimental results obtained show that the proposed system can achieve 94% accuracy in classifying goodware and DDoS malware and also accuracy of 81.8%was obtained when classifying goodware and two main malware families. [55] presented a fuzzy and fast fuzzy pattern tree methods for malware detection and categorization by using the programs' OpCodes that were transmuted into a vector space. A high degree of accuracy was achieved using a fast fuzzy pattern tree method. Dynamic analysis of malware using run-time opcodes is presented by [56]. Machine learning techniques are employed. An accuracy of 99.01% is obtained which shows that dynamic opcode analysis can be used for the detection of malware.

Also, a multi-label fuzzy clustering system for malware attack attribution is presented by [29]. Opcode frequencies were utilized as the feature space to classify different malware families. An accuracy of 94.66%, 97.56%, and 94.26% was obtained from using this classifier on identified samples from VirusShare, BIG2015 and RansomwareTracker. [19] proposed an IoT malware threat hunting using a deep recurrent neural network-based approach. Opcodes were extracted to build the datasets for malware and benign ware samples. For each opcode sample, feature vector files were created. The Long Short-Term Memory (LSTM) was used to design the deep learning structure for the detection of the IoT malware samples based on the Opcodes' sequence. Google Tensor Flow was used as a backend structure and Scikit-learn as the machine learning library for evaluation. Detection accuracy of 98% was achieved when evaluated with ARM-based IoT applications' execution codes.[2] used Convolutional Network for detection of malware in IoT and Internet of Battlefield Things (IoBT) using OpCodes. Selected Op-Code sequence was used as a feature for the classification task after which a graph of features was created for each sample. Malware classification was carried out using deep Eigenspace learning. The approach achieved an accuracy of 98.37% in malware detection and a precision rate of 98.59%. The approach also can mitigate against junk code insertion attacks.

3 Methodology

This section presents an overview of the methodology used for this study. The IoT opcode dataset used for this study was developed by researchers at the Cyber Security Laboratory of the University of Guelph. The dataset contains the program opcode for 268 different samples of goodware (benign) and 244 different samples of malware stored in separate folders. These opcodes are stored as text files. The following steps were carried out.

3.1 Feature Selection and Extraction

The IoT malware dataset is composed of 512 text files. The 512 text files will be stored as a sequence of comma-separated values (csv) for feature extraction. This is done by writing a program in Python programming language. The program extracts the vocabulary of all opcodes in the dataset. This is used to create the dictionary of words with corresponding frequency for each opcode in each of the text files. The dictionary of words (opcodes) and their frequency is exported to Microsoft Excel. There are 681 possible feature values for each of the text files. These 681 feature

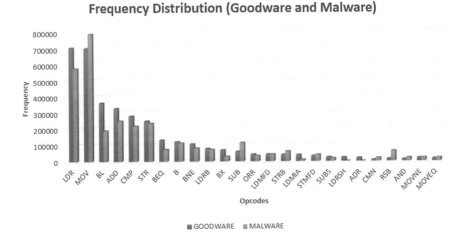

Fig. 1 Column chart showing the top 25 opcodes in the dataset

values consist of 305 unique opcodes and 376 names of the application processors embedded in each text file. Based on the new dataset created, it is observed that opcodes LDR, MOV, BL, ADD, CMP, STR have high frequency pattern in both the goodware and malware as seen in Fig. 1.

The number of feature values is large which is a bad thing as it gets to a point where more features can decrease the accuracy of a model. This is known as the curse of dimensionality. This issue can be overcome by utilizing Principal Component Analysis (PCA) algorithm. PCA algorithm helps to compress a dataset into a lower-dimensional feature subspace to retain the relevant information. This is implemented using the Python library scikit-learn.

3.2 Machine Learning Classification

The machine learning classifier is employed here for classification of malware. The five chosen classifiers are known to give excellent results when used for the classification. The classifiers to be implemented using Python library scikit-learn are listed below:

Random Forest Random Forest algorithm is a supervised classification algorithm built on an ensemble of independently trained decision trees. Random Forest grows many classification trees. They are trained using the bagging method. Feature bagging makes the random forest more robust. Feature bagging is the process by which the random forest algorithm randomly samples elements of the predictor space thereby reducing the variance of the trees at the cost of equal or higher bias.

Support Vector Machine Support Vector Machine is a discriminative classifier defined by a separating hyperplane. The objective of the SVM is to find a hyperplane in N-dimensional space (N is the number of features) that distinctly classify the data point. Hyperplanes are decision boundaries used to classify data points.

K Nearest Neighbor (KNN) It is an easy-to-implement supervised machine learning algorithm used to solve classification problems. It assumes that similar things are close to each other and do not need any training data points for model generation.

Naïve Bayes This is a probabilistic machine learning model used for classification tasks. It is based on the Bayes Theorem. They are fast and easy to implement.

Decision Tree Decision tree is used to determine a course of action. It is capable of fitting complex datasets for classification tasks. It searches for a pair of variable-value in the training set. It splits it in a way that will generate the best two-child subsets. Based on optimal splitting criteria, it creates branches and leaves. This process is called tree growing. At every node or branch, a conditional statement classifies the data point based on a fixed threshold within a specified variable thereby splitting the data. To make predictions, every new instance starts at the root node and moves along the branches until it reaches the leaf node where no further branching is possible.

4 Experimental Results

Here the results for the experiment and performance evaluation of malware detection using Random Forest (RF), K-Nearest Neighbor (KNN), Decision Tree (DT), Naïve Bayes (NB) and Support Vector Machine (SVM) is presented. To conduct this experiment, the IoT opcode dataset now transformed to a csv file will be used. The features are selected using PCA. Only the most prominent opcodes found in both the malware and goodware sample are selected, which provides a basis for the successful classification of the dataset.

The experiments were conducted on a laptop with Intel Core i7-8750H CPU of 2.20GH and 16GB of RAM. The experiments were run using Python programming language on Jupyter notebook running on Microsoft Windows 7 Professional Virtual Machine. In evaluating the efficiency of the machine learning classifiers in the detection of malware, the following criteria will be used:

True Positive (TP) implies that a malware identified as a malicious application is correct

True Negative (TN) implies that a goodware (benign) identified as a non-malicious application is actually a goodware

False Positive (FP) implies that a goodware identified as a malicious application is false

False Negative (FN) implies that malware is not identified as a malicious application.

By using the criteria above, the following metrics then be computed to quantify a given system:

Accuracy is the number of samples that is detected correctly by the classifier divided by the number of all malware and goodware application.

$$Accuracy = \frac{TP + TN}{TP + TN + FP + FN} \tag{1}$$

Precision is the ratio of predicted malware correctly labeled malware. It is defined as follows:

$$Precision = \frac{TP}{TP + FP} \tag{2}$$

Recall or detection is the ratio of malware samples correctly predicted. It is defined as follows:

$$Recall = \frac{TP + TN}{TP + FN} \tag{3}$$

F-Measure is the harmonic mean of precision and recall. It is defined as follows:

$$F - Measure = \frac{2 * TP}{2 * TP + FP + FN} \tag{4}$$

The machine learning models were trained with 409 randomly chosen malware and goodware samples and were subsequently tested with 103 goodware and malware samples. The test was done to check the accuracy of each model. Results obtained from the model is seen in Table 1 and Fig. 2 below.

From Table 1 and Fig. 2, it can be observed that changes in the number of features does not affect Random Forest and Decision Tree. Changes in the number of features has little effect on the remaining classifiers. The performance of both random forest and support vector machine classifiers across a varying number of selected features is demonstrated as seen above.

From Table 2, the Random Forest classifier has a higher training and testing time which increased as the number of features increased. Also, K-Nearest Neighbor has

Table 1 Accuracy Measure of Classifiers

No of features Selected	Classifiers				
	NB	KNN	RF	DT	SVM
10	64	98	100	99	94
20	64	97	100	99	93
30	63	95	100	99	93

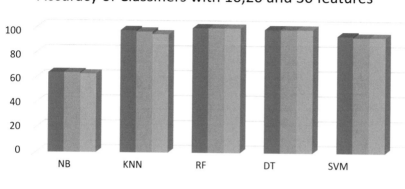

Fig. 2 Accuracy of classifiers using 10, 20 and 30 features

Table 2 Showing the training time and testing time in seconds

Classifier	No of features selected	Training time (s)	Testing time(s)
NB	10	0.002	0.001
	20	0.002	0.001
	30	0.002	0.002
KNN	10	0.002	0.006
	20	0.002	0.007
	30	0.002	0.012
RF	10	0.121	0.014
	20	0.132	0.008
	30	0.137	0.011
DT	10	0.002	0.001
	20	0.003	0.001
	30	0.003	0.001
SVM	10	0.004	0.001
	20	0.006	0.001
	30	0.006	0.001

a high testing time which increases as the number of features increases. The other classifiers have considerable low training and testing time. From Table 3, Naïve Bayes has the highest false-positive rate of 0.43 while Random Forest has the lowest false positive rate of 0.

Table 3 TPR, FPR, Precision, Recall and F-Measure for the different number of features selected

Classifier	Number of features selected	TPR	FPR	Precision	Recall	F-Measure
NB	10	0.58	0.42	0.58	0.29	0.46
	20	0.58	0.42	0.58	0.58	0.56
	30	0.57	0.43	0.57	0.29	0.45
KNN	10	0.96	0.04	0.96	0.96	0.98
	20	0.94	0.06	0.94	0.94	0.97
	30	0.92	0.08	0.92	0.91	0.95
RF	10	1.00	0	1.00	1.00	1.00
	20	1.00	0	1.00	1.00	1.00
	30	1.00	0	1.00	1.00	1.00
DT	10	0.99	0.01	0.99	0.99	0.99
	20	0.97	0.03	0.97	0.97	0.99
	30	0.97	0.03	0.97	0.97	0.99
SVM	10	0.89	0.11	0.89	0.87	0.93
	20	0.88	0.12	0.88	0.86	0.93
	30	0.88	0.12	0.88	0.86	0.93

5   Conclusion and Future Work

In this study, we have been able to implement the Random Forest, K-Nearest Neighbor, Decision Tree, Naïve Bayes, and Support Vector Machine on IoT malware dataset. Both Random Forest and Decision Tree classifiers produced high accuracy. Random Forest has the highest accuracy of 100%, Decision Tree has an accuracy of 99% while Naïve Bayes has a low accuracy of approximately 64%. There was a little change in accuracy in all classifiers except Random forest and Decision Tree when the number of features was increased. Also, the time taken for training and testing when the number of features was increased particularly for Random Forest.

For future research, the sequence of IoT malware opcodes can be examined to know if there is a specific pattern for identifying goodware and malware. Also, testing of these machine learning algorithms with a larger dataset can be explored.

References

1. S. Watson, A. Dehghantanha, Digital forensics: The missing piece of the internet of things promise. Comput. Fraud Secur. **2016**(6), 5–8 (2016). https://doi.org/10.1016/s1361-3723(15)30045-2
2. A. Azmoodeh, A. Dehghantanha, K.-K.R. Choo, Robust malware detection for internet of (battlefield) things devices using deep eigenspace learning. IEEE Trans. Sustain. Comput. **4**(1), 88–95 (2018)

3. S. Walker-Roberts, M. Hammoudeh, A. Dehghantanha, A systematic review of the availability and efficacy of countermeasures to internal threats in healthcare critical infrastructure. IEEE Access **6**, 25167–25177 (2018). https://doi.org/10.1109/ACCESS.2018.2817560
4. A. Yazdinejad, R.M. Parizi, A. Dehghantanha, H. Karimipour, G. Srivastava, M. Aledhari, Enabling drones in the internet of things with decentralized Blockchain-based security. IEEE Internet Things J. **1** (2020). https://doi.org/10.1109/jiot.2020.3015382
5. S. Nakhodchi, A. Dehghantanha, H. Karimipour, Privacy and security in smart and precision farming: A bibliometric analysis, in *Handbook of Big Data Privacy*, (Springer, Cham, 2020), pp. 305–318
6. M. Conti, A. Dehghantanha, K. Franke, S. Watson, Internet of things security and forensics: Challenges and opportunities. Futur. Gener. Comput. Syst. **78**, 544–546 (2018). https://doi.org/10.1016/j.future.2017.07.060
7. A. Yazdinejad, G. Srivastava, R.M. Parizi, A. Dehghantanha, H. Karimipour, S.R. Karizno, SLPoW: Secure and low latency proof of work protocol for Blockchain in green IoT networks, in *2020 IEEE 91st Vehicular Technology Conference (VTC2020-Spring)*, (IEEE, Antwerp, Belgium, 2020), pp. 1–5
8. A. Yazdinejad, R.M. Parizi, G. Srivastava, A. Dehghantanha, K.-K.R. Choo, Energy efficient decentralized authentication in internet of underwater things using blockchain, in *2019 IEEE Globecom Workshops (GC Wkshps)*, (IEEE Waikoloa, HI, USA, 2019), pp. 1–6
9. A. Singh, K. Click, R.M. Parizi, Q. Zhang, A. Dehghantanha, K.-K.R. Choo, Sidechain technologies in blockchain networks: An examination and state-of-the-art review. J. Netw. Comput. Appl. **149**, 102471 (2020). https://doi.org/10.1016/j.jnca.2019.102471
10. A. Yazdinejad, R.M. Parizi, A. Dehghantanha, Q. Zhang, K.-K.R. Choo, An energy-efficient SDN controller architecture for IoT networks with blockchain-based security. IEEE Trans. Serv. Comput. **13**, 625 (2020)
11. D. Połap, G. Srivastava, A. Jolfaei, R.M. Parizi, Blockchain technology and neural networks for the internet of medical things, in *IEEE INFOCOM 2020 - IEEE Conference on Computer Communications Workshops (INFOCOM WKSHPS)*, (2020), pp. 508–513. https://doi.org/10.1109/INFOCOMWKSHPS50562.2020.9162735
12. A. Yazdinejad, G. Srivastava, R.M. Parizi, A. Dehghantanha, K.-K.R. Choo, M. Aledhari, Decentralized authentication of distributed patients in hospital networks using Blockchain. IEEE J. Biomed. Heal. Inform. **24**, 2146 (2020)
13. A. Yazdinejad, R.M. Parizi, A. Bohlooli, A. Dehghantanha, K.-K.R. Choo, A high-performance framework for a network programmable packet processor using P4 and FPGA. J. Netw. Comput. Appl. **156**, 102564 (2020)
14. A. Yazdinejad, R.M. Parizi, A. Dehghantanha, K.-K.R. Choo, Blockchain-enabled authentication handover with efficient privacy protection in SDN-based 5G networks. IEEE Trans. Netw. Sci. Eng. **8**(2), 1120–1132 (2019)
15. M. Conti, T. Dargahi, A. Dehghantanha, Cyber threat intelligence: Challenges and opportunities, in *Advances in Information Security*, (Springer, Cham, 2018), pp. 1–6. https://doi.org/10.1007/978-3-319-73951-9_1
16. H. HaddadPajouh, R. Khayami, A. Dehghantanha, K.-K.R. Choo, R.M. Parizi, AI4SAFE-IoT: An AI-powered secure architecture for edge layer of internet of things. Neural Comput. Applic. **32**(20), 16119–16133 (2020). https://doi.org/10.1007/s00521-020-04772-3
17. A. Azmoodeh, A. Dehghantanha, K.-K.R. Choo, Big data and internet of things security and forensics: Challenges and opportunities, in *Handbook of Big Data and IoT Security*, (Springer, Cham, 2019), pp. 1–4. https://doi.org/10.1007/978-3-030-10543-3_1
18. H.M. Rouzbahani, H. Karimipour, A. Rahimnejad, A. Dehghantanha, G. Srivastava, Anomaly detection in cyber-physical systems using machine learning, in *Handbook of Big Data Privacy*, (Springer, Cham, 2020), pp. 219–235
19. H. HaddadPajouh, A. Dehghantanha, R. Khayami, K.-K.R. Choo, A deep recurrent neural network based approach for internet of things malware threat hunting. Futur. Gener. Comput. Syst. **85**, 88–96 (2018). https://doi.org/10.1016/j.future.2018.03.007

20. A. Yazdinejad, R.M. Parizi, A. Dehghantanha, K.-K.R. Choo, P4-to-blockchain: A secure blockchain-enabled packet parser for software defined networking. Comput. Secur. **88**, 101629 (2020). https://doi.org/10.1016/j.cose.2019.101629

21. H. HaddadPajouh, A. Dehghantanha, R.M. Parizi, M. Aledhari, H. Karimipour, A survey on Internet of Things security: Requirements, challenges, and solutions, Int. Thing. Elsevier. **14**, 100129 (2019). https://doi.org/10.1016/j.iot.2019.100129

22. H. Darabian et al., Detecting Cryptomining malware: A deep learning approach for static and dynamic analysis. J. Grid Comput. **18**, 1–11 (2020)

23. H.H. Pajouh, A. Dehghantanha, R. Khayami, K.-K.R. Choo, Intelligent OS X malware threat detection with code inspection. J. Comput. Virol. Hacking Tech. **14**(3), 213–223 (2018)

24. H. Darabian, A. Dehghantanha, S. Hashemi, S. Homayoun, K.R. Choo, An opcode-based technique for polymorphic internet of things malware detection. Concurr. Comput. Pract. Exp. **32**(6), e5173 (2020)

25. M. Zolotukhin, T. Hämäläinen, Detection of zero-day malware based on the analysis of opcode sequences, in *2014 IEEE 11th Consumer Communications and Networking Conference (CCNC)*, (IEEE Las Vegas, NV, USA, 2014), pp. 386–391

26. H. Darabian et al., A multiview learning method for malware threat hunting: Windows, IoT and android as case studies. World Wide Web **23**(2), 1241–1260 (2020)

27. N. Milosevic, A. Dehghantanha, K.-K.R. Choo, Machine learning aided android malware classification. Comput. Electr. Eng. **61**, 266–274 (2017)

28. H. Haddadpajouh, A. Azmoodeh, A. Dehghantanha, R.M. Parizi, MVFCC: A multi-view fuzzy consensus clustering model for malware threat attribution. IEEE Access **8**, 139188–139198 (2020)

29. M. Alaeiyan, A. Dehghantanha, T. Dargahi, M. Conti, S. Parsa, A multilabel fuzzy relevance clustering system for malware attack attribution in the edge layer of cyber-physical networks. ACM Trans. Cyber-Physical Syst. **4**(3), 1–22 (2020)

30. A.N. Jahromi et al., An improved two-hidden-layer extreme learning machine for malware hunting. Comput. Secur. **89**, 101655 (2020)

31. S. Homayoun et al., DRTHIS: Deep ransomware threat hunting and intelligence system at the fog layer. Futur. Gener. Comput. Syst. **90**, 94–104 (2019). https://doi.org/10.1016/j.future.2018.07.045

32. S.M. Tahsien, H. Karimipour, P. Spachos, Machine learning based solutions for security of internet of things (IoT): A survey. J. Netw. Comput. Appl. **161**, 102630 (2020)

33. M.S. Alam, S.T. Vuong, Random forest classification for detecting android malware, in *2013 IEEE International Conference on Green Computing and Communications and IEEE Internet of Things and IEEE Cyber, Physical and Social Computing*, (IEEE Beijing, China, 2013), pp. 663–669

34. M. Damshenas, A. Dehghantanha, K.-K.R. Choo, R. Mahmud, M0droid: An android behavioral-based malware detection model. J. Inf. Priv. Secur. **11**(3), 141–157 (2015)

35. T. Lu, S. Hou, A two-layered malware detection model based on permission for android, in *2018 IEEE International Conference on Computer and Communication Engineering Technology (CCET)*, (IEEE Beijing, China, 2018), pp. 239–243

36. W. Wang, M. Zhao, J. Wang, Effective android malware detection with a hybrid model based on deep autoencoder and convolutional neural network. J. Ambient. Intell. Humaniz. Comput. **10**(8), 3035–3043 (2019)

37. E. Karbab, M. Debbabi, A. Derhab, D. Mouheb, MalDozer: Automatic framework for android malware detection using deep learning. Digit. Investig. **24**, S48–S59 (2018)

38. W. Li, Z. Wang, J. Cai, S. Cheng, An android malware detection approach using weight-adjusted deep learning, in *2018 International Conference on Computing, Networking and Communications (ICNC)*, (IEEE Maui, HI, USA, 2018), pp. 437–441

39. A. Pektaş, T. Acarman, Deep learning for effective android malware detection using API call graph embeddings. Soft. Comput. **24**(2), 1027–1043 (2020)

40. T. Kim, B. Kang, M. Rho, S. Sezer, E.G. Im, A multimodal deep learning method for android malware detection using various features. IEEE Trans. Inf. Forensic. Secur. **14**(3), 773–788 (2018)
41. Y.-S. Yen, H.-M. Sun, An android mutation malware detection based on deep learning using visualization of importance from codes. Microelectron. Reliab. **93**, 109–114 (2019)
42. M. Kruczkowski, E.N. Szynkiewicz, Support vector machine for malware analysis and classification, in *2014 IEEE/WIC/ACM International Joint Conferences on Web Intelligence (WI) and Intelligent Agent Technologies (IAT)*, vol. 2, (IEEE Warsaw, Poland, 2014), pp. 415–420
43. R.S. Pirscoveanu, S.S. Hansen, T.M.T. Larsen, M. Stevanovic, J.M. Pedersen, A. Czech, Analysis of malware behavior: Type classification using machine learning, in *2015 International Conference on Cyber Situational Awareness, Data Analytics and Assessment (CyberSA)*, (IEEE London, United Kingdom, 2015), pp. 1–7
44. M. Imran, M.T. Afzal, M.A. Qadir, Using hidden markov model for dynamic malware analysis: First impressions, in *2015 12th International Conference on Fuzzy Systems and Knowledge Discovery FSKD*, (2015), pp. 816–821. https://doi.org/10.1109/FSKD.2015.7382048
45. A. Makandar, A. Patrot, Malware analysis and classification using artificial neural network, in *International Confererence on Trends in Automation Communications and Computing Technology I-TACT 2015*, (2016), p. 7492653. https://doi.org/10.1109/ITACT.2015.7492653
46. M.L. Bernardi, M. Cimitile, F. Martinelli, F. Mercaldo, A fuzzy-based process mining approach for dynamic malware detection, in *2017 IEEE International Conference on Fuzzy Systems (FUZZ-IEEE)*, (IEEE Naples, Italy, 2017), pp. 1–8
47. H. Hashemi, A. Azmoodeh, A. Hamzeh, S. Hashemi, Graph embedding as a new approach for unknown malware detection. J. Comput. Virol. Hacking Tech. **13**(3), 153–166 (2017)
48. Y. Ding, W. Dai, S. Yan, Y. Zhang, Control flow-based opcode behavior analysis for malware detection. Comput. Secur. **44**, 65–74 (2014)
49. Q. Jerome, K. Allix, R. State and T. Engel, Using opcode-sequences to detect malicious Android applications, in *2014 IEEE International Conference on Communications (ICC)*, Sydney, (IEEE, Sydney, Australia, 2014), pp. 914–919
50. B. Kang, S.Y. Yerima, K. McLaughlin, S. Sezer, N-opcode analysis for android malware classification and categorization, in *2016 International Conference on Cyber Security and Protection of Digital Services (Cyber Security)*, (IEEE, London, UK, 2016), pp. 1–7
51. J. Baldwin, A. Dehghantanha, Leveraging Support Vector Machine for Opcode Density Based Detection of Crypto-Ransomware, in *Cyber Threat Intelligence. Advances in Information Security*, (Cham, Springer, 2018), pp. 107–136
52. H. Zhang, X. Xiao, F. Mercaldo, S. Ni, F. Martinelli, A.K. Sangaiah, Classification of ransomware families with machine learning based on N-gram of opcodes. Futur. Gener. Comput. Syst. **90**, 211–221 (2019)
53. G. Canfora, F. Mercaldo, C.A. Visaggio, Mobile malware detection using op-code frequency histograms, in *2015 12th International Joint Conference on e-Business and Telecommunications (ICETE)*, vol. 4, (IEEE Colmar, France, 2015), pp. 27–38
54. J. Su, V.D. Vasconcellos, S. Prasad, S. Daniele, Y. Feng, K. Sakurai, Lightweight classification of IoT malware based on image recognition, in *2018 IEEE 42Nd Annual Computer Software and Applications Conference (COMPSAC)*, vol. 2, (IEEE Tokyo, Japan, 2018), pp. 664–669
55. A. Azmoodeh, A. Dehghantanha, R.M. Parizi, H. Karimipour, E. Modiri, D.E. Newton, Fuzzy pattern tree for edge malware detection and categorization in IoT zero trust distributed computing view project naive-Bayesian-based model for interoperability among heterogeneous Systems in Intelligent Buildings View project fuzzy pattern tree for. Art. J. Syst. Arch. **97**, 1 (2019)
56. D. Carlin, P. O'Kane, S. Sezer, Dynamic analysis of malware using run-time opcodes, in *Data Analytics and Decision Support for Cybersecurity*, (Springer, Cham, 2017), pp. 99–125

Mac OS X Malware Detection with Supervised Machine Learning Algorithms

Samira Eisaloo Gharghasheh and Shahrzad Hadayeghparast

1 Introduction

During recent years, a considerable utilization of digital system for all aspects of life has been recorded which has turned these systems as gold mines for cyberattackers [1–5]. A potentially harmful software to both networks and computers is called malware [6–9]. The information security faces main threats by the significant increase in the malware variants. Each day malicious software can deliver [10] up to 360,000 malware samples, which are new or changed, based on Kaspersky's 2017 Security Report [11]. The number of OS X malware developed by cyber threat actors is on the increase because of their fast adoption rate as well as the rise in the use of Mac OS X devices [12]. For instance, according to the McAfee Labs Threats Report [11], the Mac OS X malware increased by 744% from 2015 to 2016.

The detection of new types of malware as well as classifying the unknown malware is not possible by the anti-virus systems which use signatures [10]. Machine learning is used as an effective tool for malware detection to deal with the ever-increasing complexity and diversity of malware [13–16]. As a result, machine learning techniques are adopted to detect new and unknown malware. To the best of the author's knowledge and the study presented in [12] there is a lot of research carried out for detecting Android [17, 18] and windows [19–23] malware using machine learning algorithms, but only a few numbers of studies have considered OS X malware.

S. E. Gharghasheh (✉) · S. Hadayeghparast
School of Computer Science, University of Guelph, Guelph, ON, Canada
e-mail: samira@cybersciencelab.org; shadayeg@uoguelph.ca

© Springer Nature Switzerland AG 2021
K.-K. R. Choo, A. Dehghantanha (eds.), *Handbook of Big Data Analytics and Forensics*, https://doi.org/10.1007/978-3-030-74753-4_13

In this paper, different machine learning algorithms are adopted in order to detect OS X malware. The various machine learning techniques used fall into five main categories of Decision Tree, Support Vector Machine (SVM), K-Nearest Neighbors (KNN), Ensemble and Logistic Regression. A number of new features based on libraries are developed and added to the original OS X dataset in order to improve the performance of the machine learning algorithms in detection. Accordingly, one feature is created for each library and the corresponding feature value is 1 or 0 whether the library is called or not called respectively. The performance of machine learning algorithms can be compared based on performance metrics and ROC cure. Three of these performance metrics presented in Eqs. 1, 2 and 3 are used in this paper [24].

$$True\ positive\ rate\ (TPR) = \frac{TP}{TP + FP} \tag{1}$$

$$False\ negtive\ rate\ (FNR) = \frac{FN}{FN + TN} \tag{2}$$

$$Accuracy = \frac{TP + TN}{FN + TP + FP + TN} \tag{3}$$

Where TPR is the rate that the classifier correctly predicts benign. FPR is the rate that the malware is classified incorrectly as benign. Accuracy measures the ratio that a classifier correctly detects benign and malware samples. True Positive (TP) is the number of correctly classified benign. True Negative (TN) is the number of correctly classified malware. False Positive (FP) is the number of benign incorrectly classified as malware. False Negative (FN) is the number of malwares incorrectly classified as benign.

The receiver operating curve known as the ROC curve is a useful means for the evaluation of machine learning algorithms as well as making a comparison between them. ROC curve is the measurement of TPR versus FPR [25]. The Area Under the ROC Curve known as AUC is also an important measure for comparing the performance of machine learning algorithms. Prediction is evaluated based on different AUC values as follows [24]: the values of 1.0, 0.9, 0.8, 0.7, 0.6, 0.5 and under 0.5 represent perfect, excellent, good, mediocre, poor, random and poor respectively.

This paper is organized as follows. Section 2 presents the related work in this field. Section 3 provides the methodology. Results and discussion are presented in Sect. 4. Finally, this paper is concluded in Sect. 5.

2 Related Work

There are many studies conducted in the field of detecting malware using machine learning and algorithms, a number of which is presented in this section [26–32]. The anomaly-based approach is used as an alternative work for detecting malware in [24]. Among the machine learning algorithms used in this paper, the k-nearest neighbor algorithm achieved the TPR of 84.57% in detecting the latest Android malware. Also, a multi-layer perceptron achieved 93.03% TRP, while Random Forest attends the TPR of 99.97 on the MalGenome dataset. Kernel-based Support Vector Machine was used in [12] for detecting OS X malware. In addition, a new weighting method, which uses the frequency of library calling, was presented as a novel approach. The false alarm rate of 3.9% and detection accuracy above 91% was achieved using the kernel based SVM and the weighting measure. An incremental malware detection system is presented in [10] for detecting new malware as well as classifying malware families. In the aforesaid study, new malware families are detected by adopting the Shared Nearest Neighbor (SNN) clustering algorithm. In addition, the authors could achieve an accuracy of 86.7% and 98.9% in the detection of new malware and the classification of the unknown malware. Two strategies are proposed in [17] for statistical analysis of android malware based on machine learning. In the first approach, the permission-based classification models achieved the F-measure of 89%. Moreover, the second approach, which is the source code-based classification, attained an F-score of 95.1%. A dynamic analysis strategy based on machine learning is proposed in [25]. A large-cale dataset on a real Android device was gathered that includes 4816 and 1866 benign and malicious applications, respectively. For this purpose, combinations of random forest classifier and Conformal Prediction machine learning algorithms are adopted in the aforesaid study. In order for unknown applications to be classified, Graph Community Algorithms and machine learning are used in [33] for a better combination of different Multi-scanner Antivirus detections. It is noteworthy that this study has attained an F1-score of more than 0.87. The authors have used five machine learning algorithms in [34] and obtained 91.43% accuracy on ransomware dataset. Moreover, the binary classification accuracy of 99.3% and the average F1-measure of 99% is achieved in this study. Both supervised and unsupervised machine learning algorithms along with utilizing opcode frequency for feature vector are adopted for malware detection in [35]. A Two-hidden-layered Extreme Learning Machine (TELM) has built in [19] to automatically detect malware. They achieved an accuracy of 99.65% in IoT dataset malware detection. In [21], the researchers have used deep learning methods for cryptomining malware detection. Their analysis was both static and dynamic which achieved an accuracy of 95% and 99% respectively. A novel taxonomy of crypto-ransomware features and an analysis had presented in [36]. The behavior of crypto-ransomware and efficient detection methods had been analyzed in their paper based on the Cyber Kill Chain (CKC) framework. In order to detect malware in the IoT dataset [37] had proposed a fuzzy and a fast fuzzy pattern tree. The fuzzy pattern tree achieved an accuracy of 99.83% while the fast fuzzy model achieved an accuracy of 100%. Authors in [38] for detecting malicious applications in IoT environment, has used sequential pattern mining method. In polymorphic IoT malware detection they achieved an accuracy of 99%. For ransomware detection, a Deep Ransomware Threat Hunting and Intelligence System (DRTHIS) had been proposed in [20]. In ransomware classification, they achieved a TPR of 97.2% and F-measure of 99.6%. Authors in [39] had used the Recurrent Neural Network (RNN) to detect IoT malware. By applying 2-layer neurons, they achieved the highest accuracy of 98.18%. By transmuting the device's Operational Code (OpCode) into a vector space, [40] had presented deep learning for malware detection on the Internet of Battlefield Things (IoBT). Their accuracy and precision were as follows 99.68% and 98.59% respectively. In [22], to detecting ransomware attacks, they put forward a new machine learning method. Their method had monitored the patterns of energy consumption in applications to classify them as ransomware or non-malicious application. Their method had achieved a precision rate of 89.19% and a detection rate of 95.65%.

3   Methodology

This section presents the research methodology which is illustrated in Fig. 1. First, the OS X dataset is described, and its features are defined in the dataset subsection. In the second step, the preprocessing procedure is explained in detail. Next, the technique used for feature selection is presented. Finally, the machine learning algorithms used in this paper are described.

3.1   Dataset

In this paper, the OS X dataset is taken from [25], which includes 459 benign and 152 malware. This dataset is nonbiased due to the fact that the number of good wares is three times the number of malwares. The description of features in this dataset is provided in Table 1.

Fig. 1 The workflow of the research

Table 1 The features of OS X Dataset [12]

No. Feature	Feature name	Description	Value type
1	Ncmds	Number of commands of each sample	Integer
2	sizeofcmds	Size of commands of each sample	Integer
3	noloadcmd	Number of commands which sample will loaded during execution	Integer
4	rebase_size	Define size of the rebase information	Integer
5	bind_size	Define size of the information which will be bind during execution	Integer
6	lazy_bind_size	Define size of the information which will be bind during execution	Integer
7	export_size	Define the size of the lazy binding information	Integer
8	nsyms	Define the number of symbol table entries	Integer
9	strsize	Define string table size in bytes	Integer
10	LoadDYLIB	Define number of DYLIB which called and load for executing of malware	Integer
11	DYLIBnames	Define names of loaded DYLIB	Nominal
12	Segments	Number of total segments which consist in each sample	Integer
13	SectionsTEXT	Number text segments which consist in each sample	Integer
14	SectionsData	Number data segments which consist in each sample	Integer

3.2 Preprocessing

Preprocessing of the dataset is one of the important parts of the malware detection. In this paper, a thorough preprocessing is performed on the OS X dataset in MATLAB which includes the following steps:

- Dropping the duplicate samples
- Correcting the missing values
- Redetermining the number of loaded DYLIB
- Feature scaling
- Creating new features for called libraries

The definition of features are taken from apple developer guideline [12].

3.2.1 Dropping the Duplicate Samples

In this step, the duplicate examples are identified and removed from the OS X dataset. The duplicates are specified according to the similar names found in the first column of the dataset "name". Consequently, 23 examples are removed from the dataset.

3.2.2 Correcting the Missing Values

Three type of missing values are detected in this dataset including:

NANs
Presence of string data types instead of numeric values
Combination of string and numeric data types in one cell

The first two types of missing values are corrected by replacing the corresponding cell with the previous entry in the same column. However, for the third type, the string part is separated from the numeric part and removed.

3.2.3 Redetermining the Number of Loaded DYLIB

The number of called and loaded libraries whose names are presented in "DYLIB-names" is demonstrated in its previous column. However, these numbers are recalculated, and incorrect numbers are observed in some of the examples. Consequently, those incorrect values are replaced with the true amount.

3.2.4 Feature Scaling

The range of features used in this dataset varies significantly. Consequently, the features with larger numbers have more impact on the classifier model. For achieving better accuracy, the feature values Scaled into the range of [0,1].

3.2.5 Creating New Features for Called Libraries

Creating new features based on libraries presented in "DYLIBnames" column is proposed in this paper. For this purpose, first, the list of all libraries called and loaded is extracted from "DYLIBnames" column. Then, one feature is created for each library. The total number of 589 features are added to the dataset accordingly. It is noteworthy that feature values in the new columns are 1 if the library is used by each sample, otherwise, it is zero.

3.3 Feature Selection

For determining the features which contribute the most to classification results, feature selection techniques are adopted. In this paper, the Chi-Square technique is used for scoring features using Eqs. 4 and 5 as follows [12]:

$$X^2(t, c) = \frac{N \times (AD - CB)^2}{((A + C) \times (B + D) \times (A + B) \times (C + D))} \quad (4)$$

$$X^2_{avg}(t) = \Pr(c_i) X^2(t, c_i) \quad (5)$$

Where score function is demonstrated by $X^2_{avg}(t)$. Where D is the frequency without the occurrence of c or t. The times c occurs without t is denoted by C. The frequency that t happens without c is expressed by B. The frequency of occurrence of t and c simultaneously is denoted by A. the sample size is demonstrated by N.

3.4 Machine Learning Algorithms

Twenty-one machine learning algorithms from five main categories of Decision Tree, SVM, KNN, Ensemble and Logistic Regression are used in this paper as shown in Table 2.

4 Results and Discussion

The results of adopting machine learning algorithms for the detection of OS X malware are presented and analyzed in three case studies. In case one, the libraries used by malware and benign wares are ignored; consequently, the dataset includes 13 features excluding "DYLIBnames". The impact of considering each library as a new feature is investigated in the next case study. In the last case study, the most important features are selected, and the resulted evaluation metrics and run-time are compared with the second case study.

4.1 Case 1

Table 3 presents the performance metrics for applying machine learning algorithms. The highest accuracy of 90.5% is obtained by Subspace KNN which is an Ensemble classifier. According to this table, Ensemble classifiers had better performance compare to other classifiers in terms of accuracy. Regarding TPR percentage, SVM Coarse gaussian is ranked first with 99%. Consequently, the aforesaid classifier is the number one for identifying benign correctly. Concerning FNR, RUSBoosted trees which is an Ensemble classifier headed the first with 9%. Therefore, this classifier has the best performance in detecting malware correctly.

Table 2 Machine learning algorithms

Classifier name	Decision Tree	SVM	KNN	Ensemble	Logistic Regression
	Complex tree	Linear	Fine	Boosted trees	Logistic regression
	Medium tree	Quadratic	Medium	Bagged trees	
	Simple tree	Cubic	Coarse	Subspace discriminant	
		Fine gaussian	Cosine	Subspace KNN	
		Medium gaussian	Cubic weighted	RUSBoosted trees	
		Coarse gaussian			

Table 3 Performance metrics for Case 1

Classifier	Accuracy%	TPR%	FNR%
Decision tree			
Complex tree	88.6	92	22
Medium tree	87.6	91	22
Simple tree	80.5	86	36
SVM			
Linear	80.3	95	68
Quadratic	84.2	89	31
Cubic	83.1	88	32
Fine gaussian	87.1	94	37
Medium gaussian	81.9	95	60
Coarse gaussian	77.8	99	91
KNN			
Fine	89.3	92	19
Medium	83.6	92	43
Coarse	77.5	96	81
Cosine	84.7	92	40
Cubic	83.1	91	43
Weighted	86.6	92	30
Ensemble			
Boosted trees	89.7	93	22
Bagged trees	89.8	94	23
Subspace discriminant	79.3	97	77
Subspace KNN	90.5	94	22
RUSBoosted trees	89.2	89	9
Logistic regression			
Logistic regression	81.7	92	50

The ROC Curves for the classifiers having the highest accuracy in each group including Complex Tree, Fine gaussian, KNN Fine, Subspace KNN and Logistic Regression are shown in Fig. 2. As it is shown in this figure, Subspace KNN, which had the highest accuracy among the classifiers, has also the highest AUC of 0.92.

4.2 Case 2

Table 4 presents the evaluation metrics for machine learning algorithms applied to the OS X dataset having all the library features. Concerning accuracy, in the first place was Subspace KNN with 94.7%. Coarse gaussian achieved TPR of 100%, although it has the worst performance among classifiers in detecting malware. However, this is not a good result because the aforesaid algorithm considered almost all software as benign. As it is shown in this table, two KNN and two Ensemble classifiers have the least FNR of 9%.

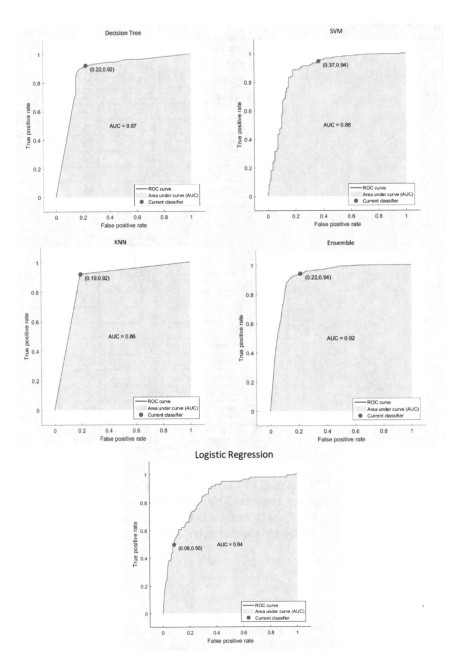

Fig. 2 ROC Curves for Case 1

Table 4 Performance metrics for Case 2

Classifier	Accuracy%	TPR%	FNR%
Decision tree			
Complex tree	91.7	95	19
Medium tree	91.4	95	21
Simple tree	86.6	98	49
SVM			
Linear	93.6	97	17
Quadratic	92.7	95	15
Cubic	92.9	95	15
Fine gaussian	91.5	97	27
Medium gaussian	89.5	98	39
Coarse gaussian	78	100	94
KNN			
Fine	92.7	93	9
Medium	89.8	93	22
Coarse	85.8	99	56
Cosine	91.9	94	14
Cubic	90.2	93	19
Weighted	92.5	93	9
Ensemble			
Boosted trees	93.2	96	17
Bagged trees	94.2	96	12
Subspace discriminant	93.7	98	19
Subspace KNN	94.7	96	9
RUSBoosted trees	93.4	94	9
Logistic regression			
Logistic regression	81.2	82	21

The ROC Curves for Case 2 are shown in Fig. 3. The highest AUC belongs to the Ensemble classifier called Subspace KNN with 0.99. Linear SVM, Complex tree, Fine KNN and Logistic Regression are in the next place with 0.97, 0.93, 0.92 and 0.81 AUCs respectively.

4.3 Case 3

Bagged trees as an Ensemble classifier has the highest accuracy with 92.2%. Coarse gaussian has classified all benign correctly having TPR 100% while it shows poor performance in detecting malware. RUSBoosted trees are in first place in detecting malware in FNR of 9%.

Bagged trees, weighted KNN, Linear SVM, complex Decision tree and Logistic Regression have the AUC of 0.97, 0.96, 0.94, 0.84 and 0.80 respectively. As it is shown in this figure the highest AUC belongs to the Ensemble classifier.

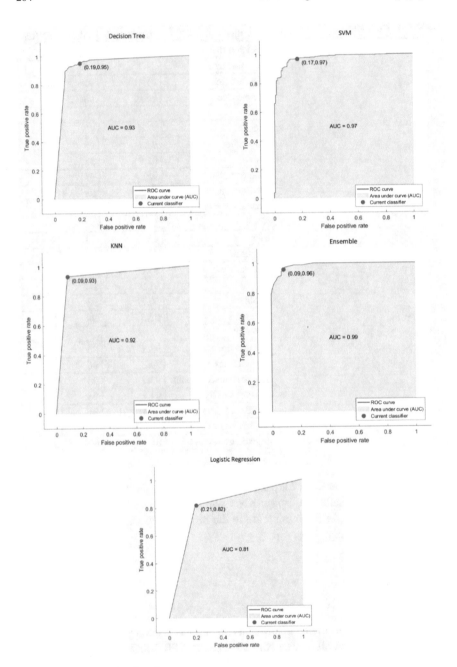

Fig. 3 ROC Curves for Case 2

Table 5 Best Accuracy in each Case study

	Case 1	Case 2	Case 3
Classifier	Ensemble	Ensemble	Ensemble
Algorithm	Subspace KNN	Subspace KNN	Bagged trees
Accuracy	90.5	94.7	92.2

According to the results obtained in Case 1, 2, 3 considering the libraries as independent features improve the accuracy of machine learning classifiers. The accuracy increased by about 4% and 2% re-spectively in comparison with the Case 1 in which libraries were ig-nored. The highest accuracy in each Case study is also demonstrated in Table 5. Moreover, Table 5 shows that Ensemble classifiers have the best performance on OS X dataset compared to the other classi-fiers used in this paper.

5 Conclusion

In this paper, various machine learning algorithms from five main categories of Decision Tree, SVM, KNN, Ensemble and Logistic Regression were used for detecting Mac OS malware. Performance metrics of accuracy, TPR and, FNR, as well as ROC curve, were used for evaluating their performance. The Ensemble classifies demonstrated the best performance among other classifiers having the highest accuracies. In addition, a novel technique of developing new features based on library calls was adopted. Three case studies were performed to investigate the impact of the aforesaid new features. In the first case, the library calls were ignored and the highest accuracy of 90.5% was achieved by Subspace KNN as an Ensemble classifier. In Case 2, 589 features for library calls were added which resulted in the accuracy of 94.7% by the same algorithm, demonstrating an increase in accuracy by 4%. In addition, performance metrics and AUC of almost all classifiers were increased. This result shows that taking into account the new features for library calls improves the performance of machine learning algorithms in detection. Also, in case 3, feature selection based on the Chi-Square technique was employed which selected half of the features. The best accuracy decreased by about 2% which shows that the amount of rising in the accuracy is almost linearly dependent on the number of library features. Finally, our future work would be adopting deep learning for detecting Mac OS malware.

References

1. S. Nakhodchi, A. Dehghantanha, H. Karimipour, Privacy and security in smart and precision farming: A bibliometric analysis, in *Handbook of Big Data Privacy*, (Springer, Cham, 2020), pp. 305–318
2. S. Walker-Roberts, M. Hammoudeh, A. Dehghantanha, A systematic review of the availability and efficacy of countermeasures to internal threats in healthcare critical infrastructure. IEEE Access **6**, 25167–25177 (2018 March). https://doi.org/10.1109/ACCESS.2018.2817560
3. H.M. Rouzbahani, H. Karimipour, A. Dehghantanha, R.M. Parizi, Blockchain applications in power systems: A bibliometric analysis, in *Blockchain Cybersecurity, Trust and Privacy*, ed. by K.-K. R. Choo, A. Dehghantanha, R. M. Parizi, vol. 79, (Springer, Cham)
4. M. Conti, A. Dehghantanha, K. Franke, S. Watson, Internet of things security and forensics: Challenges and opportunities. Futur. Gener. Comput. Syst. **78**, 544–546 (2018). https://doi.org/10.1016/j.future.2017.07.060
5. A. Yazdinejad, R.M. Parizi, A. Dehghantanha, K.-K.R. Choo, P4-to-blockchain: A secure blockchain-enabled packet parser for software defined networking. Comput. Secur. **88**, 101629 (2020). https://doi.org/10.1016/j.cose.2019.101629
6. I. Santos, J. Devesa, F. Brezo, J. Nieves, P.G. Bringas, Opem: A static-dynamic approach for machine-learning-based malware detection, in *International Joint Conference CISIS'12-ICEUTE 12-SOCO 12 Special Sessions*, (Springer, Berlin, Heidelberg, 2013), pp. 271–280
7. H. Hashemi, A. Azmoodeh, A. Hamzeh, S. Hashemi, Graph embedding as a new approach for unknown malware detection. J. Comput. Virol. Hacking Tech. **13**(3), 153–166 (2017)
8. A. Azmoodeh, A. Dehghantanha, Big data and privacy: Challenges and opportunities, in *Handbook of Big Data Privacy*, (Springer, Cham, 2020), pp. 1–5. https://doi.org/10.1007/978-3-030-38557-6_1
9. A. Azmoodeh, A. Dehghantanha, K.-K.R. Choo, Big data and internet of things security and forensics: Challenges and opportunities, in *Handbook of Big Data and IoT Security*, (Springer, Cham, 2019), pp. 1–4. https://doi.org/10.1007/978-3-030-10543-3_1
10. L. Liu, B. Wang, B. Yu, Q. Zhong, Automatic malware classification and new malware detection using machine learning. Front. Inf. Technol. Electron. Eng. **18**(9), 1336–1347 (2017)
11. McAfee, *McAfee Labs Threats Report: April 2017*, no. April (2017), p. 49
12. H.H. Pajouh, A. Dehghantanha, R. Khayami, K.-K.R. Choo, Intelligent OS X malware threat detection with code inspection. J. Comput. Virol. Hacking Tech. **14**(3), 213–223 (2018)
13. A. Demontis et al., Yes, machine learning can be more secure! a case study on android malware detection. IEEE Trans. Depend. Secur. Comput. **16**(4), 711–724 (2017)
14. M. Saharkhizan, A. Azmoodeh, A. Dehghantanha, K.-K.R. Choo, R.M. Parizi, An ensemble of deep recurrent neural networks for detecting IoT cyber attacks using network traffic. IEEE Internet Things J. **7**(9), 8852–8859 (2020). https://doi.org/10.1109/jiot.2020.2996425
15. M. Saharkhizan, A. Azmoodeh, H. HaddadPajouh, A. Dehghantanha, R.M. Parizi, G. Srivastava, A hybrid deep generative local metric learning method for intrusion detection, in *Handbook of Big Data Privacy*, (Springer International Publishing, Cham, 2020), pp. 343–357. https://doi.org/10.1007/978-3-030-38557-6_16
16. A. Yazdinejad, A. Bohlooli, K. Jamshidi, Efficient design and hardware implementation of the OpenFlow v1.3 switch on the Virtex-6 FPGA ML605. J. Supercomput. **74**(3), 1299 (2018). https://doi.org/10.1007/s11227-017-2175-7
17. N. Milosevic, A. Dehghantanha, K.-K.R. Choo, Machine learning aided android malware classification. Comput. Electr. Eng. **61**, 266–274 (2017)
18. M. Damshenas, A. Dehghantanha, K.-K.R. Choo, R. Mahmud, M0droid: An android behavioral-based malware detection model. J. Inf. Priv. Secur. **11**(3), 141–157 (2015)
19. A.N. Jahromi et al., An improved two-hidden-layer extreme learning machine for malware hunting. Comput. Secur. **89**, 101655 (2020)

20. S. Homayoun et al., DRTHIS: Deep ransomware threat hunting and intelligence system at the fog layer. Futur. Gener. Comput. Syst. **90**, 94–104 (2019). https://doi.org/10.1016/j.future.2018.07.045
21. H. Darabian et al., Detecting Cryptomining malware: A deep learning approach for static and dynamic analysis. J. Grid Comput. **18**, 1–11 (2020)
22. A. Azmoodeh, A. Dehghantanha, M. Conti, K.-K.R. Choo, Detecting crypto-ransomware in IoT networks based on energy consumption footprint. J. Ambient. Intell. Humaniz. Comput. **9**(4), 1141–1152 (2018)
23. S. Homayoun, A. Dehghantanha, M. Ahmadzadeh, S. Hashemi, R. Khayami, Know abnormal, find evil: Frequent pattern mining for ransomware threat hunting and intelligence. IEEE Trans. Emerg. Top. Comput. **8**, 341 (2017)
24. F.A. Narudin, A. Feizollah, N.B. Anuar, A. Gani, Evaluation of machine learning classifiers for mobile malware detection. Soft. Comput. **20**(1), 343–357 (2016)
25. H. Papadopoulos, N. Georgiou, C. Eliades, A. Konstantinidis, Android malware detection with unbiased confidence guarantees. Neurocomputing **280**, 3–12 (2018)
26. A. Yazdinejad, H. HaddadPajouh, A. Dehghantanha, R.M. Parizi, G. Srivastava, M.-Y. Chen, *Cryptocurrency malware hunting: A deep recurrent neural network, in Applied Soft Computing*, vol **96**, (Elsevier, 2020), p. 106630
27. A. Yazdinejad, R.M. Parizi, G. Srivastava, A. Dehghantanha, K.K.R. Choo, Energy efficient decentralized authentication in internet of underwater things using blockchain, in *2019 IEEE Globecom Workshops (GC Wkshps)*, (IEEE, 2019), pp. 1–6
28. M. Aledhari, R. Razzak, R.M. Parizi, F. Saeed, Federated learning: A survey on enabling technologies, protocols, and applications. IEEE Access **8**, 140699–140725 (2020). https://doi.org/10.1109/ACCESS.2020.3013541
29. A. Yazdinejad, R.M. Parizi, A. Dehghantanha, H. Karimipour, G. Srivastava, M. Aledhari, Enabling drones in the internet of things with decentralized Blockchain-based security. IEEE Internet Things J., 1 (2020). https://doi.org/10.1109/jiot.2020.3015382
30. V. Mothukuri, R.M. Parizi, S. Pouriyeh, Y. Huang, A. Dehghantanha, G. Srivastava, A survey on security and privacy of federated learning. Futur. Gener. Comput. Syst. **115**, 619 (2020)
31. R.M. Parizi, S. Homayoun, A. Yazdinejad, A. Dehghantanha, K.-K.R. Choo, *Integrating Privacy Enhancing Techniques into Blockchains Using Sidechains* (2019). https://doi.org/10.1109/CCECE.2019.8861821
32. A. Yazdinejad, R.M. Parizi, A. Dehghantanha, G. Srivastava, S. Mohan, A.M. Rababah, Cost optimization of secure routing with untrusted devices in software defined networking. J. Parallel Distrib. Comput. **143**, 36 (2020)
33. I. Martín, J.A. Hernández, S. de los Santos, Machine-learning based analysis and classification of android malware signatures. Futur. Gener. Comput. Syst. **97**, 295–305 (2019)
34. H. Zhang, X. Xiao, F. Mercaldo, S. Ni, F. Martinelli, A.K. Sangaiah, Classification of ransomware families with machine learning based on N-gram of opcodes. Futur. Gener. Comput. Syst. **90**, 211–221 (2019)
35. H. Aghakhani, G. Fabio, M. Francesco, L. Martina, O. Stefano, B. Davide, V. Giovanni, K. Christopher, When malware is Packin'Heat; limits of machine learning classifiers based on static analysis features, in *Network and Distributed Systems Security (NDSS) Symposium 2020*. (2020)
36. T. Dargahi, A. Dehghantanha, P.N. Bahrami, M. Conti, G. Bianchi, L. Benedetto, A cyber-kill-chain based taxonomy of crypto-ransomware features. J. Comput. Virol. Hacking Tech. **15**(4), 277–305 (2019)
37. E.M. Dovom, A. Azmoodeh, A. Dehghantanha, D.E. Newton, R.M. Parizi, H. Karimipour, Fuzzy pattern tree for edge malware detection and categorization in IoT. J. Syst. Archit. **97**, 1–7 (2019)
38. H. Darabian, A. Dehghantanha, S. Hashemi, S. Homayoun, K.R. Choo, An opcode-based technique for polymorphic internet of things malware detection. Concurr. Comput. Pract. Exp. **32**(6), e5173 (2020)

39. H. HaddadPajouh, A. Dehghantanha, R. Khayami, K.-K.R. Choo, A deep recurrent neural network based approach for internet of things malware threat hunting. Futur. Gener. Comput. Syst. **85**, 88–96 (2018). https://doi.org/10.1016/j.future.2018.03.007
40. A. Azmoodeh, A. Dehghantanha, K.-K.R. Choo, Robust malware detection for internet of (battlefield) things devices using deep eigenspace learning. IEEE Trans. Sustain. Comput. **4**(1), 88–95 (2018)

Machine Learning for OSX Malware Detection

Alex Chenxingyu Chen and Kenneth Wulff

1 Introduction

Attackers pay more attention to the macOS platform today because its global market share, which jumped to 14.37% in 2019 [1] has been rising steadily since that data was captured. According to Wikipedia, there has been a further increase of 3.58% since the start of 2019 [2]. This surge in user and corporate adoption of the macOS operating system has come with its challenges because cybercriminals have gradually shifted some of their attention to the macOS operating system and have developed various malware that use numerous techniques, tactics and procedures to attack victims and avoid detection [3].

The software security company, McAfee, reports that ransomware activity has been resurgent this year. McAfee estimates that new ransomware grew by 118% and that criminals have adopted new tactics and code innovations for attack execution and systems evasion. McAfee says its systems witnessed 504 new threats every minute in the first quarter of 2019 and predicts that the trend is likely to continue.

Other estimates from McAfee show that total macOS malware growth significantly increased from around 200,000 in the fourth quarter of 2016 to approximately 425,000 in the third quarter of 2018 [4]. These numbers are for detected malware only. The bigger concern for ordinary users and corporate clients alike is the damage some of these undetected malwares may be causing. For example, in 2019, a new malware family called CookieMiner was uncovered to be aiming at Apple users and sending code to ultimately steal user wallets and authorizations. It was embedded as a library in macOS and was used to send the stolen coins and credentials to an xmrig server. Another new exploit kit, known as Spelevo, was also exposed in the

A. C. Chen (✉) · K. Wulff
School of Computer Science, University of Guelph, Guelph, ON, Canada
e-mail: cchen22@uoguelph.ca; kwulff@uoguelph.ca

© Springer Nature Switzerland AG 2021
K.-K. R. Choo, A. Dehghantanha (eds.), *Handbook of Big Data Analytics and Forensics*, https://doi.org/10.1007/978-3-030-74753-4_14

first quarter of 2019. Spelevo exploited a vulnerability in the Adobe Flash Player to drop the GootKit Trojan. It exploited this code execution vulnerability and allowed the remote attacker to execute arbitrary code [5].

According to Kaspersky Labs, the notion that 'there are no threats or at least no serious threats for the macOS operating system' and by extension, its users is a gross misconception. Kaspersky says the threat landscape is changing because the 'popularity of the Apple platform is growing.' According to them, nearly six million phishing attacks targeting macOS users were detected during the first half of 2019. Of the six million, 11.80% were directed at corporate users. Kaspersky state that the number of phishing attacks that 'make use of the Apple brand name grows by 30 – 40% per year'. According to the company, the number of malicious and potentially unwanted programs have been on the ascendency since 2012. The company estimates that malicious attacks targeting macOS users exceeded four million in 2018, and reported an estimated 1.8 million such attacks in the first half of 2019 [6].

The proliferation of these malicious software supports the notion that traditional anti-virus or anti-malware programs have not been successful at mitigating the spread of these malicious programs [7–11]. This has therefore called for the need to develop new tools, techniques and strategies to help address this growing global anathema [12, 13].

Many solutions have been proposed to deal with this growing concern of malware attacks [14–17]. Some of the solutions have involved the use of frameworks that use both static and dynamic analysis techniques to help detect malware [18]. The limitation for some of these frameworks is that it is unable to detect malware that alter their signature, behaviour or code from time to time [19].

Machine learning, which has been used to detect malware in computers running other operating systems such as Windows [15, 20], Linux [21] and Android [22, 23] offer a reliable and promising solution for macOS users. Much work, however, has not been undertaken in this area even though there has been a jump in macOS usage worldwide. It is because of this that we decided to carry out this research to help determine how different machine learning algorithms can help to deal with this growing problem and to also contribute to the body of knowledge as researchers find increasingly reliable methods to combat these risks.

In our research, we evaluate five different machine learning algorithms on their ability to effectively detect macOS malware. Our objective is to determine which of the five machine learning algorithms produces the best results.

To understand this research, the reader needs to have knowledge about machine learning algorithms. Machine learning is a data analysis technique that uses artificial intelligence to help systems develop the ability to automatically learn and improve from experience without being explicitly programmed [24]. Machine learning uses data and different techniques to build a mathematical model to predict or classify unseen data.

In this paper, we will be using the supervised learning method to conduct our experiments. Supervised learning algorithms build a blueprint or model based on input and given output. The input data and the resulting output consist of training data. When a machine learning model has been trained, the data used for the training can be used to examine the performance of the model used. The evaluation or training data will be given to the model as input. The output that is churned out can then be compared with real data. When two outputs match, the model is said to have successfully classified the data [25].

The next section of this paper is dedicated to the literary review, which addresses others' work on this topic. Section 3 gives a breakdown of the methodology used for the research. The main parts of Sect. 3 are data cleaning, feature transformation and feature selection. The results are presented in Sect. 4. Finally, we talk about the conclusion and what could be done in the future in the last section.

2 Literature Review

Malware detection has become a significant security consideration for many computer systems today [26–30]. Malware detection techniques that work based on signatures do an excellent job of detecting previously discovered malware but are unable to detect polymorphic pieces of code that evolve or mutate due to changes in their signature(s). Because of the development of this kind of malicious code engineering, the accuracy of these detectors cannot be relied upon as they generate a lot of false-positives and false-negatives which ultimately affect the efficacy of systems tasked with the responsibility of uncovering these attacks and protecting the computer infrastructures from harm and malicious intent [16, 31–33]. Researchers have, therefore, conducted experiments to try and find an alternative solution that can still detect these malicious pieces of code even after they change their signatures.

Using malicious executable linkable files, Kakavand et al. [21] deployed machine learning strategies to successfully demonstrate a classification accuracy of 97% in Linux, in 2014. They achieved this by dynamically extracting system calls using a system call tracer named Strace to build a classification model that could efficiently identify best feature sets and group them into benign and malware specimens.

To add to the insufficient malware detection solutions available on Linux systems Asmitha and Vinod proposed a novel model which used machine learning to detect malicious Executable Linkable Files. They achieved this by dynamically extracting system calls using a tracer called Strace.

Their experiment produced a malware classification accuracy of 97%. Using various machine learning methods, they were able to achieve an accuracy rate of 97.3% with Random Forest. AdaboostMI(J48) and J48 produced an accuracy rates of 96.70% and 94.91% respectively. They also indicate that an increase in the feature length significantly impact the classification accuracy because of the noise that begin to appear in these machine learning models.

Using static analysis, Mohsen et al., in 2018, used real-world malware and benign apps to show an average precision rate of 79.08% and a true-positive rate of 67.00% for SVM, and an 80.50% accuracy rate for KNN with a true-positive rate of 80.00% on Android [34]. Their research compared two machine learning algorithms, namely, SVM and KNN, to group Android applications as benign or malware. The above results were based on measuring the accuracy and true-positive rates for detecting malware on Android.

In measuring the gap between in-the-lab and in-the-wild validation scenarios, Kevin et al., used empirical assessment of machine learning-based methods to detect malware on Android. Their research relied on a set of features built from control flow graphs of over 50,000 Android applications collected from varied sources. Their research showed a stark contrast between results achieved in a lab setting and results achieved in the real world. Using a tenfold cross-validation assessment, they revealed that the experiments revealed a poor performance overall in detecting malware [35].

They show that even though variations of goodware/malware ratio and classification algorithms produce the same results, increasing the features lead to a significant drop in the ability of these machine learning methods to detect malware on a large scale [36]. Their research helped to identify several parameters that impact the performance of malware at scale in the real world. In 2019, three researchers from India achieved almost 100% accuracy for malware detection in windows using five machine learning algorithms. The machine learning techniques they used are Random Forest, Logistic Model Tree (LMT), NBT (Naïve Bayes Tree), J48 Graft and REPTree. Their research was based on the occurrence of operation codes (opcodes) or instruction syllables to improve the accurate detection of advanced unknown malware [37]. The researchers used the Fisher Score method for their feature selection to help them to overcome code obfuscation used by advanced malware to evade anti-malware tools. The Fisher Score method statistically helped them to solve the maximum likelihood equation problem to uncover previously unknown malicious pieces of code.

In their paper titled review of machine learning methods for windows malware detection, Saima and Dushyant went beyond the traditional signature-based malware detection methods which failed at detecting unknown malware executable files and presented a simple but efficient malware detection mechanism that was able to detect benign and malicious executable files by extracting features from the Portable Executable (PE) headers. Their research used the static analysis approach, and it was able to detect malware before installation of the executable files.

The accomplished this by utilizing various machine learning techniques such as Support Vector (SV), Decision Tree (DT), Random Forest (RF) and Gaussian Naïve Bays (GNB) classifiers. Their experiment showed that of all the methods used, the Random Forest classifier had the highest accuracy rate of 98.63%. The next highest was the Ada Boost Classiffier (ADB) which achieved an accuracy rate of 97.26%. The lowest was the BernoulliNaive Bays Classifier (BNB) with an accuracy rate of 67.12% [38].

In an effort to help stem the spread of malware due to the widespread adoption of Internet of Things (IoT) [39–47], Ayush and Teng from the National University of Singapore presented a distributed modular solution called EDIMA to help detect IoT malware activity on large scale networks such as enterprise ISP networks. EDIMA uses machine learning algorithms to classify traffic packets using a vector database, policy module and an optional sub-sampling packet module.

The feature values for benign and malicious data utilized telnet, http post and http get categories. Using this approach, they were able to show rates of 77.78%, 88.8% and 94.44% for Gaussian Naïve Bayes, Random Forest and k-NN respectively. Their model shows that the test beds were mostly hardware based, which has a lot of limitations. As indicated in their conclusion, a software-based test bed would be more ideal [47].

In June 2019, Duy-Phuc Pham et al. published their research results, where they used a hybrid malware analysis framework that combined both static and dynamic analysis to help analysts detect malicious software on macOS. The framework featured a kernel hooking module, which used various analysis techniques including system calls and other service invocations specific to macOS to help detect malware. They were able to use this technique through argument monitoring, process tracing, anti-analysis detection and mitigation to uncover 71 unknown macOS malware variants that have evaded current detection methods. Their research showed that even though 85% of the samples detected were adware, 49% of them belonged to the trojan/backdoor family and utilized techniques such as sleep probing to evade detection [48]. The limitation for this framework was its ability to still detect these malicious software codes after their signatures or modus operandi had changed.

Using application library calls, Pajouh et al., used a supervised machine learning prototype and applied kernel base SVM in conjunction with a revolutionary calculation to detect malware and benign applications on macOS operating systems with a 91% accuracy and a 3.9% false alarm rate [49] in 2018. The researchers went a step further and used Synthetic Minority Over-sampling Technique (SMOTE) to create three synthetic datasets with distinctive distributions according to refined variations of the dataset to investigate the impact different sample sizes will have on malware detection accuracy ratios. The SMOTE technique achieved over 96% detection accuracy with a false alarm rate of below 4%.

The goal of this research is to show that the Decision Tree algorithm offers the best solution in detecting malware with a true-positive rate of 97.83% and a false positive rate of 3.62%. We also seek to use this research to contribute to the development and improvement of machine learning strategies for macOS malware analyses and to add to the body of knowledge in this critical and growing space.

3   Methodology

The methodology section outlines the data processing and training workflow used for the experiment. Figure 1 provides an overview of the various steps that we followed to generate the results we discuss in a later section of this research paper.

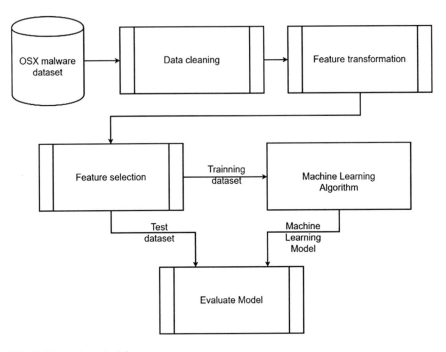

Fig. 1 Research methodology

3.1 Environment Setup and Dataset Download

The operating system we used to perform the analysis was a Windows 10 Pro, version 18,362, with an Intel Core i7 9700k CPU. The CPU speed clock speed was 3.6GHz, and it was also fitted with 32GB of RAM.

We used the Anaconda 2019.10 Windows version to set up the Python 3.7 environment and jupyter lab.

Anaconda is an all-in-one installer, which includes all the necessary python packages for data science.

We used the OSX malware dataset from the Cyber Science lab [49]. This dataset includes 461 benign samples and 152 malicious samples.

3.2 Data Cleaning

First, we converted the dataset format from XLSX to CSV because the read_xlsx function did not work on our research platform. It is normal to get data that are incomplete, noisy or inconsistent. This is called dirty data. Inside the OSX malware

Table 1 Feature list of OSX malware dataset

Feature name	Type
ncmds	int
sizeofcmds	int
noloadcmd	int
rebase_size	int
bind_size	int
lazy_bind_size	int
export_size	int
nsyms	int
strsize	int
LoadDYLIB	int
DYLIBnames	string
Segments	int
SectionsTEXT	int
SectionsData	int

dataset, we had one record that lacked a feature. Also, a couple of records had inconsistent data [50].

For the missing feature, which was a numerical NaN, we replaced the missing value with the mean of the other values. Some records had a mixture of hexadecimal and decimal numbers, which we resolved by converting the hexadecimals to decimal numbers. After cleaning the data, we started to perform the feature engineering process on the dataset. This included feature transformation and feature selection, details of which have been provided below.

3.3   Feature Transformation

Table 1 shows the feature list of the OSX malware dataset. The most feature type is an integer. However, the DYLIBnames column had strings, which are the list of system libraries that were used by benign or malicious applications. All the machine learning algorithms that we chose required numerical data for training and testing. We, therefore, needed to encode the textual data. The technique we chose to encode the texture data was one-hot encoding [51]. This technique converts the textual data based on features. If one library is present in a data point, that feature will be set to one.

After converting all the data into integers, we normalized the data by using min-max scaling, which was necessary for improving the rendition of the machine learning algorithm [52]. The scaling limited all data into a range of (0,1). The scaling was needed for the SVM algorithm as it could avoid the features that have higher numeric fields dominating other features that have a smaller range.

3.4 Feature Selection

We used the statistical test chi-squared [53], to test the relationships between each feature and the class label. Then we chose the best 40 features for Support Vector Machine and Naïve Bayes. We decided not to apply feature selection for the Decision Tree, Stochastic Gradient Descent and Logistic Regression algorithms because those algorithms can handle all features quickly. Another reason we did not use the feature selection for the Decision Tree, Stochastic Gradient Descent and Logistic Regression algorithms was also because when we filtered out some features, the accuracy rate for the Decision Tree algorithm dropped. Nevertheless, when we applied the feature selection on Naïve Bayes, the accuracy increased considerably. After completing the feature engineering process, we started training the algorithms.

3.5 Machine Learning Classifier Phase

The machine learning algorithm results are presented in this section. To validate the machine learning accuracy, we used the cross-validation technique [54] with ten runs, and computed the final results by averaging the outputs from the cross-validations. Below is a brief description of the various machine learning algorithms that we used for our experiment.

Decision Tree It is a supervised machine learning algorithm. It uses a bunch of rules or guidelines to classify a data point. It divides the data into little subsets to form a tree structure. The tree will have both decision and leaf nodes. Decision nodes represent the rules, and the leaf nodes represent a class label [55]. We trained and tested our machine learning model by using Gini impurity and Information gain.

SVM SVM is also a supervised machine learning algorithm. Furthermore, it can be used for categorization and regression. SVM creates a line named decision line between different class labels. SVM will try to maximize the margin around the decision line [56]. Because solving SVM problems involve quadratic programming, we decided to filter out some features to improve the training performance.

SGD SGD is a supervised machine learning algorithm. It is one of the three common gradient descents. SGD is an algorithm that is simple but has a high-efficiency rate. It is also very suitable for large-scale learning. For example, as the number of OSX malware increases, high efficiency could reduce the time required for training and let researchers apply the model faster [57].

Naïve Bayes Naïve Bayes is a machine learning algorithm based on the Bayes theorem. It is good at classifying tasks. It assumes that all features are independent of each other, and therefore, a change in one feature does not influence other features [58].

LR Logistic Regression is a supervised machine learning that is also used for classification tasks. It has been widely used to identify spam emails and fraudulent online transactions [59].

4 Results and Discussion

In this section, we calculate the performance of the malware detection algorithms. The accuracy evaluation used one hundred cross-validation runs with the OSX malware dataset as input. We then computed the mean of the one hundred results as the final accuracy for that particular machine learning model. And used the confusion matrix [60] to evaluate the details of the algorithm performance. The following were adopted from the confusion matrix to help evaluate the classifier.

$$\text{True Positive Rate (TPR)} := \frac{True\ positive}{True\ positive + False\ negative} \tag{1}$$

$$\text{False Positive Rate (FPR)} := \frac{False\ positive}{True\ negative + False\ positive} \tag{2}$$

$$\text{Precision} := \frac{True\ positive}{True\ positive + False\ positive} \tag{3}$$

$$\text{Recall} := \frac{True\ positive}{True\ positive + False\ negative} \tag{4}$$

$$\text{F1 Score} := 2 \times \frac{Precision \times Recall}{Precision + Recall} \tag{5}$$

The true-positive is the number of applications that the benign program marks as normal. True-negative is the number of apps that the malware characterizes as malicious. False-positive is the number of normal applications that are marked as malicious. False-negative represents the number of malicious apps that were undetected. The true-positive rate shows the percentage of successful malware detections. The false-positive rate shows the possibility of the classifier marking a healthy macOS program as malware. High precision means that not many benign applications were predicted as malware. High recall means that the algorithm could correctly detect or predict most malware.

First, we tested and compared the decision tree performance between Gini impurity and Information gain. The Gini impurity criterion gave a 92% cross-validation

Table 2 Decision tree performance comparison

Criterion	TPR (%)	FPR (%)	Precision	Recall	F1 score
Information gain	97.83	3.62	0.978	0.920	0.938
Gini impurity	84.00	5.97	0.840	0.840	0.840

Table 3 Evaluation results for malware detection

Classifier	Accuracy (%)	TPR (%)	FPR (%)	Precision	Recall	F1 score
Decision tree	92.78	97.83	3.62	0.978	0.920	0.938
SGD	91.77	93.02	7.09	0.930	0.800	0.860
LR	89.77	89.58	5.15	0.896	0.860	0.878
SVM	88.33	96.97	1.19	0.970	0.640	0.771
Naïve Bayes	87.54	96.97	11.92	0.970	0.640	0.771

accuracy, and the Information gain produced a 93% cross-validation correctness. Although the two methods have similar accuracy in Table 2, Information gain has other attributes that make it a better choice than Gini impurity.

Second, we tested the Decision Tree algorithm and reported an accuracy rate of 92.78%, with a false-positive ratio of 3.62%. After that, we tested the Stochastic Gradient Descent algorithm. The SGD algorithm also used all the feature columns. With the default parameters, it reported a 91.77% accuracy. Its true-positive rate is lower than the Decision Tree. It also has a higher false-positive ratio compared to the Decision Tree algorithm. The Logistic Regression algorithm, which we tested with default parameters and all the features, reported an overall accuracy of 89.77%. The Support Vector Machine algorithm also used the default configuration. With 100 cross-validations, it reported an accuracy rate of 88.33%, with a slightly lower true-positive rate than the Decision Tree algorithm but with a better false-positive rate. After SVM, we tested Naïve Bayes with default parameters and all the feature columns. When we applied the feature selection on Naïve Bayes with 40 features, we had an 87.54% accuracy ratio and a 96.97% true-positive rate and an 11.92% false-positive rate. Naïve Bayes has an excellent positive rate similar to SVM, but with very high false positives. A summary of the results is shown in Table 3.

5   Conclusion and Future Work

This research outlined an assessment of the ability of machine learning algorithms to detect macOS malware. The algorithms we used were the Decision Tree, Support Vector Machine, Gaussian Naïve Bayes, Stochastic Gradient Descent and Logistic Regression. According to the results summarized in the last section, the Decision Tree has the highest accuracy ratio. Although the Decision Tree has a higher false-positive rate than the Support Vector Machine algorithm, it is insignificant when you consider the damage that can be caused by these malicious software codes. Future works can compare more machine learning algorithms to find an algorithm that has higher accuracy and lower false-positive rates than what has been reported in this research. Additionally, future works could gather more data points in order to get a more accurate and well-trained model.

References

1. Statscounter, Desktop Operating System Market Share Worldwide, StatCounter Global Stats, 2019. [Online]. Available: https://gs.statcounter.com/os-market-share/desktop/worldwide/#monthly-200901-201909. [Accessed: 11-Dec-2019].
2. Wikipedia, *Usage Share of Operating Systems in Europe, Wikipedia, 2014* (2019). https://en.wikipedia.org/wiki/Usage_share_of_operating_systems. Accessed 13 Dec 2019, p. 2019
3. McAfee, *McAfee Labs Threats Report: April 2017*, no. April (2017), p. 49
4. S. Watson, A. Dehghantanha, Digital forensics: the missing piece of the Internet of Things promise. Comput. Fraud Secur. **2016**(6), 5–8 (2016). https://doi.org/10.1016/s1361-3723(15)30045-2
5. C. Beek et al., *Mcafee Labs Threats Report*, Technical report (McAfee, St. Clara, 2017)
6. M. Kuzin, T. Shcherbakova, T. Sidorina, V. Kamluk, *Threats to macOS users | Securelist, Securelist by Kaspersky*, 2019. [Online]. Available: https://securelist.com/threats-to-macos-users/93116/. (Accessed: 13-Dec-2019).
7. P.N. Bahrami, A. Dehghantanha, T. Dargahi, R.M. Parizi, K.-K.R. Choo, H.H.S. Javadi, Cyber kill chain-based taxonomy of advanced persistent threat actors: analogy of tactics, techniques, and procedures. J. Inf. Process. Syst. **15**(4), 865–889 (2019)
8. M. Conti, T. Dargahi, A. Dehghantanha, Cyber threat intelligence: challenges and opportunities, in *Advances in Information Security*, (Springer, Cham, 2018), pp. 1–6. https://doi.org/10.1007/978-3-319-73951-9_1
9. M. Conti, A. Dehghantanha, K. Franke, S. Watson, Internet of things security and forensics: Challenges and opportunities. Futur. Gener. Comput. Syst. **78**, 544–546 (2018). https://doi.org/10.1016/j.future.2017.07.060
10. A. Azmoodeh, A. Dehghantanha, K.-K.R. Choo, Big data and internet of things security and forensics: Challenges and opportunities, in *Handbook of Big Data and IoT Security*, (Springer, Cham, 2019), pp. 1–4. https://doi.org/10.1007/978-3-030-10543-3_1
11. A. Yazdinejad, R.M. Parizi, A. Dehghantanha, K.-K.R. Choo, P4-to-blockchain: A secure blockchain-enabled packet parser for software defined networking. Comput. Secur. **88**, 101629 (2020). https://doi.org/10.1016/j.cose.2019.101629
12. T. Dargahi, A. Dehghantanha, P.N. Bahrami, M. Conti, G. Bianchi, L. Benedetto, A cyber-kill-chain based taxonomy of crypto-ransomware features. J. Comput. Virol. Hacking Tech. **15**(4), 277–305 (2019)
13. A. Yazdinejad, A. Bohlooli, K. Jamshidi, Efficient design and hardware implementation of the OpenFlow v1.3 switch on the Virtex-6 FPGA ML605. J. Supercomput. **74**(3), 1299 (2018). https://doi.org/10.1007/s11227-017-2175-7
14. S. Homayoun et al., Deep dive into ransomware threat hunting and intelligence at fog layer. Futur. Gener. Comput. Syst. **90**(Jan 19), 94–104 (2018)
15. S. Homayoun, A. Dehghantanha, M. Ahmadzadeh, S. Hashemi, R. Khayami, Know abnormal, find evil: frequent pattern mining for ransomware threat hunting and intelligence. IEEE Trans. Emerg. Top. Comput. **8**, 341 (2017)
16. A.N. Jahromi et al., An improved two-hidden-layer extreme learning machine for malware hunting. Comput. Secur. **89**, 101655 (2020)
17. A.N. Jahromi, S. Hashemi, A. Dehghantanha, R.M. Parizi, K.-K.R. Choo, An enhanced stacked LSTM method with no random initialization for malware threat hunting in safety and time-

critical systems. IEEE Trans. Emerg. Top. Comput. Intell. **4**(5), 630–640 (2020). https://doi.org/10.1109/tetci.2019.2910243

18. H. Darabian et al., Detecting cryptomining malware: a deep learning approach for static and dynamic analysis. J. Grid Comput. **18**, 1–11 (2020)

19. H. HaddadPajouh, R. Khayami, A. Dehghantanha, K.-K.R. Choo, R.M. Parizi, AI4SAFE-IoT: an AI-powered secure architecture for edge layer of Internet of things. Neural Comput. Applic. **32**(20), 16119–16133 (2020). https://doi.org/10.1007/s00521-020-04772-3

20. H. Hashemi, A. Azmoodeh, A. Hamzeh, S. Hashemi, Graph embedding as a new approach for unknown malware detection. J. Comput. Virol. Hacking Tech. **13**(3), 153–166 (2017)

21. K.A. Asmitha, P. Vinod, A machine learning approach for linux malware detection, in *Proceedings of the 2014 International Conference on Issues and Challenges in Intelligent Computing Techniques, ICICT 2014*, (2014), pp. 825–830. https://doi.org/10.1109/ICICICT.2014.6781387

22. N. Milosevic, A. Dehghantanha, K.-K.R. Choo, Machine learning aided android malware classification. Comput. Electr. Eng. **61**, 266–274 (2017)

23. K. Shaerpour, A. Dehghantanha, R. Mahmod, Trends in android malware detection. J. Digit. Forensic Secur. Law **8**(3), 2 (2013)

24. Expert.ai Team, "What is Machine Learning? A definition - Expert System | Expert.ai, Expert System, 2019. [Online]. Available: https://www.expert.ai/blog/machine-learning-definition/. [Accessed: 13-Dec-2019].

25. Wikipedia, *Supervised Learning – Wikipedia*. Wikipedia (2019). https://en.wikipedia.org/wiki/Supervised_learning. Accessed 13 Dec 2019

26. A. Azmoodeh, A. Dehghantanha, M. Conti, K.-K.R. Choo, Detecting crypto-ransomware in IoT networks based on energy consumption footprint. J. Ambient. Intell. Humaniz. Comput. **9**(4), 1141–1152 (2018)

27. A. Azmoodeh, A. Dehghantanha, K.-K.R. Choo, Robust malware detection for internet of (battlefield) things devices using deep eigenspace learning. IEEE Trans. Sustain. Comput. **4**(1), 88–95 (2018)

28. H. Haddadpajouh, A. Azmoodeh, A. Dehghantanha, R.M. Parizi, MVFCC: a multi-view fuzzy consensus clustering model for malware threat attribution. IEEE Access **8**, 139188–139198 (2020)

29. H. Darabian et al., A multiview learning method for malware threat hunting: windows, IoT and android as case studies. World Wide Web **23**(2), 1241–1260 (2020)

30. M. Saharkhizan, A. Azmoodeh, A. Dehghantanha, K.-K.R. Choo, R.M. Parizi, An Ensemble of Deep Recurrent Neural Networks for detecting IoT cyber attacks using network traffic. IEEE Internet Things J. **7**(9), 8852–8859 (2020). https://doi.org/10.1109/jiot.2020.2996425

31. K. Kosmidis, C. Kalloniatis, "Machine learning and images for malware detection and classification, in ACM International Conference Proceeding Series, 2017, vol. Part F132523, pp. 1–93.

32. M. Saharkhizan, A. Azmoodeh, H. HaddadPajouh, A. Dehghantanha, R.M. Parizi, G. Srivastava, A hybrid deep generative local metric learning method for intrusion detection, in *Handbook of Big Data Privacy*, (Springer, Cham, 2020), pp. 343–357. https://doi.org/10.1007/978-3-030-38557-6_16

33. A. Azmoodeh, A. Dehghantanha, R.M. Parizi, S. Hashemi, B. Gharabaghi, G. Srivastava, Active spectral botnet detection based on eigenvalue weighting, in *Handbook of Big Data Privacy*, (Springer, Cham, 2020), pp. 385–397. https://doi.org/10.1007/978-3-030-38557-6_19

34. M. Kakavand, M. Dabbagh, and A. Dehghantanha, Application of machine learning algorithms for android malware detection, in ACM International Conference Proceeding Series, 2018, pp. 32–36.

35. K. Allix, T.F. Bissyandé, Q. Jérome, J. Klein, R. State, Y. Le Traon, Empirical assessment of machine learningbased malware detectors for android: Measuring the gap between in-the-lab and in-the-wild validation scenarios. Empir. Softw. Eng. **21**, 183. https://doi.org/10.1007/s10664-014-9352-6

36. A. Yazdinejad, R. M. Parizi, A. Dehghantanha, and K. K. R. Choo, Blockchain-enabled authentication handover with efficient privacy protection in SDN-based 5G networks, IEEE Trans. Netw. Sci. Eng., pp. 1–1, May 2019.

37. S. Sharma, C.R. Krishna, S.K. Sahay, Detection of advanced malware by machine learning techniques, in *Soft Computing: Theories and Applications*, (Springer, Singapore, 2019), pp. 333–342

38. S. Naz and D. K. Singh, Review of Machine Learning Methods for Windows Malware Detection, in 2019 10th International Conference on Computing, Communication and Networking Technologies, ICCCNT 2019, 2019.

39. A. Yazdinejad, R.M. Parizi, G. Srivastava, A. Dehghantanha, K.K.R. Choo, Energy efficient decentralized authentication in internet of underwater things using blockchain, in 2019 IEEE Globecom Workshops, GC Wkshps 2019 - Proceedings, 2019.

40. A. Yazdinejad, H. HaddadPajouh, A. Dehghantanha, R.M. Parizi, G. Srivastava, M.-Y. Chen, *Cryptocurrency Malware Hunting: A Deep Recurrent Neural Network Approach*, vol 96 (Elsevier, 2020)

41. M. Aledhari, R. Razzak, R.M. Parizi, F. Saeed, Federated learning: A survey on enabling technologies, protocols, and applications. IEEE Access **8**, 140699–140725 (2020). https://doi.org/10.1109/ACCESS.2020.3013541

42. A. Yazdinejad, R.M. Parizi, A. Dehghantanha, H. Karimipour, G. Srivastava, M. Aledhari, Enabling drones in the internet of things with decentralized Blockchain-based security. IEEE Internet Things J., 1 (2020). https://doi.org/10.1109/jiot.2020.3015382

43. V. Mothukuri, R.M. Parizi, S. Pouriyeh, Y. Huang, A. Dehghantanha, G. Srivastava, A survey on security and privacy of federated learning. Futur. Gener. Comput. Syst. **115**, 619 (2020)

44. A. Yazdinejad, R.M. Parizi, A. Dehghantanha, G. Srivastava, S. Mohan, A.M. Rababah, Cost optimization of secure routing with untrusted devices in software defined networking. J. Parallel Distrib. Comput. **143**, 36 (2020)

45. Q. Chen, G. Srivastava, R.M. Parizi, M. Aloqaily, I. Al Ridhawi, An incentive-aware blockchain-based solution for internet of fake media things. Inf. Process. Manag. **57**, 102370 (2020). https://doi.org/10.1016/j.ipm.2020.102370

46. A. Yazdinejad, A. Bohlooli, K. Jamshidi, Performance improvement and hardware implementation of Open Flow switch using FPGA, in 2019 IEEE 5th Conference on Knowledge Based Engineering and Innovation, KBEI 2019, 2019, pp. 515–520.

47. A. Kumar and T. J. Lim, EDIMA: Early Detection of IoT Malware Network Activity Using Machine Learning Techniques, in IEEE 5th World Forum on Internet of Things, WF-IoT 2019 - Conference Proceedings, 2019, pp. 289–294.

48. D.-P. Pham, D.-L. Vu, F. Massacci, Mac-A-Mal: macOS malware analysis framework resistant to anti evasion techniques. J. Comput. Virol. Hacking Tech. **15**(4), 249–257 (2019)

49. H.H. Pajouh, A. Dehghantanha, R. Khayami, K.-K.R. Choo, Intelligent OS X malware threat detection with code inspection. J. Comput. Virol. Hacking Tech. **14**(3), 213–223 (2018)

50. Wikipedia, *Dirty Data – Wikipedia*. Wikipedia (2019). https://en.wikipedia.org/wiki/Dirty_data. Accessed 13 Dec 2019

51. Wikipedia, *One-Hot – Wikipedia*. Wikipedia (2019). https://en.wikipedia.org/wiki/One-hot. Accessed 13 Dec 2019

52. Wikipedia, *Feature Scaling – Wikipedia*. Wikipedia (2019). https://en.wikipedia.org/wiki/Feature_scaling. Accessed 13 Dec 2019

53. Wikipedia, Chi-Squared Test – Wikipedia, *Wikipedia* (2019). https://en.wikipedia.org/wiki/Chi-squared_test. Accessed 13 Dec 2019

54. A. S. 44, *Cross Validation in Machine Learning – GeeksforGeeks*. GeeksforGeeks (2017). https://www.geeksforgeeks.org/cross-validation-machine-learning/. Accessed 13 Dec 2019

55. J. Stoldt, T. Uwe Trapp, T.C. Sehra, *Decision Trees Explained Easily – Chirag Sehra – Medium*, medium.com (2018). https://medium.com/@chiragsehra42/decision-trees-explained-easily-28f23241248. Accessed 11 Dec 2019

56. L. Schultebraucks, *Introduction to Support Vector Machines, Available.* medium.com (2017).   https://medium.com/@LSchultebraucks/introduction-to-support-vector-machines-9f8161ae2fcb. Accessed 11 Dec 2019

57. Scikit Learn, *1.5. Stochastic Gradient Descent – Scikit-Learn 0.22 Documentation.* scikit-learn.org (2019). https://scikit-learn.org/stable/modules/sgd.html. Accessed 11 Dec 2019

58. Machinelearningplus.com, How Naive Bayes Algorithm Works ? (with example and full code), Machinelearningplus.com, 2018. [Online]. Available: https://www.machinelearningplus.com/predictive-modeling/how-naive-bayes-algorithm-works-with-example-and-full-code/. Accessed 11 Dec 2019

59. A. Pant, *Introduction to Logistic Regression – Towards Data Science.* towards-datascience.com (2019). https://towardsdatascience.com/introduction-to-logistic-regression-66248243c148. Accessed 11 Dec 2019

60. S. Narkhede, *Understanding Confusion Matrix – Towards Data Science.* towardsdatascience.com (2018). https://towardsdatascience.com/understanding-confusion-matrix-a9ad42dcfd62. Accessed 13 Dec 2019

Hybrid Analysis on Credit Card Fraud Detection Using Machine Learning Techniques

Akansha Handa, Yash Dhawan, and Prabhat Semwal

1 Introduction

Most of the population prefers using a credit card over cash or debit. A credit card is a convenient payment tool that is accepted worldwide. The number of credit cards in circulation is phenomenal and the number of users reaches up to 679 million as of 2018 [1]. With E-commerce becoming more and more common, there has been a rise in credit card fraud. Every year, millions of people fall victim to credit card fraud which results in millions of losses in the economy. In 2018, $24.26 Billion was lost due to payment card fraud worldwide [2]. Therefore, preserving the security, safety and privacy of financial services is necessary [3, 4]. The credit fraud can take place in a number of ways like card lost – someone finds and uses it to perform an online transaction, card details leaked out in public through shoulder surfing, social engineering attack wherein the person gives out information and through numerous other techniques [5, 6]. Other forms of attack include hijacking credit card data from online payment forms, skimming and data breaches. One such recent major breach has been the Capital One (Financial institute) breach which impacted 6 million Canadian people as well as over 100 million in the US. The following breach exposed customers PII (Personal Identifiable Information), customer credit limits, payment histories and SIN numbers [7]. Detecting fraud for any financial institute is an utmost priority which cannot be achieved by traditional tools and techniques. Many financial institutes have now migrated to machine learning and deep learning to handle credit card frauds more efficiently. In this paper, we are implementing a novel Ensemble machine learning technique that combines several base models in order to produce one optimal predictive model. Ensemble machine learning reduces

A. Handa (✉) · Y. Dhawan · P. Semwal
School of Computer Science, University of Guelph, Guelph, ON, Canada
e-mail: ahanda@uoguelph.ca; ydhawan@uoguelph.ca; psemwal@uoguelph.ca

© Springer Nature Switzerland AG 2021
K.-K. R. Choo, A. Dehghantanha (eds.), *Handbook of Big Data Analytics and Forensics*, https://doi.org/10.1007/978-3-030-74753-4_15

variance, noise and bias as well as increases the accuracy of the model [8]. To get optimal results we balance our dataset using sampling technique – SMOTE (Synthetic Minority Oversampling Technique) and Cluster centroid. Basic machine learning analysis is performed on the balanced dataset using Supervised Learning (Logistic Regression, Random Forest and Support Vector Machine), Unsupervised Learning (K means) and Deep Neural Networks. Following which a hybrid analysis is performed on the Supervised learning algorithm using a bagging ensemble technique to achieve a strong predictive model.

Section 2 of this paper contains a literature review on similar papers from recent years. Section 3 details the methodology used in this work, Sect. 4 contains the results of our experiments, and in Sect. 5 we draw conclusions and suggest future work. References for our work are provided at the very end of the paper.

2 Related Works

In earlier studies, many computer scientists have proposed various approaches [9–17] to resolve fraud transaction detection problem using different techniques of machine learning. The remainder of this section describes the different fraud detection models implemented by past researchers.

In 2017, J. O. Awoyemi performed a comparative study on three different machine learning algorithms (Naive Bayes, K-nearest neighbor and logistic regression) based on TPR, TNR, FPR and FNR rates metrics. The dataset was sampled using a hybrid sampling approach where stepwise addition and subtraction was done on both positive and negative data points. As a result, two sets of dataset distribution (10:90) and (34:64) were generated. The highest accuracy of 97.92% was observed with KNN on 34:64 data distribution and KNN. Awoyemi et al. [18] Similarly, In another comparative study, the performance of various supervised machine learning algorithms was analyzed based on the accuracy, TPR, FRP, specificity and G-mean. All the supervised classifiers were compared with a super classifier which was implemented using the stacking method of ensemble learning technique. The overall result indicated the significant performance of the stacking classifier in detecting the fraud transaction with 95.2% accuracy. Furthermore, the stacking classifier achieved 95% precision and 95% recall [19]. In the paper [20], a real-time fraud detection system having three different models was introduced. In this system, the Fraud detection model was responsible for detecting the fraud transaction and passing the recognized fraud transaction to the other two models: API and Data warehouse. The supervised classifiers like Logistic regression, Naive Bayes and Super Vector Machine were selected through a literature study and their performance in addressing the four different fraud patterns in the sampled dataset. The highest accuracy of 91% was achieved with Super Vector Machine. In 2019, used different sampling techniques to deal with imbalanced data problem in case of credit card datasets. As a result, different datasets were generated: undersampled, oversampled and one with synthetic data (using KNN). The Logistic regression,

Decision tree and extreme gradient boosting machine learning algorithms were trained with all three sets of datasets. The highest accuracy of 99.75% was achieved for Logistic regression algorithms on the undersampled dataset [21].

Many researchers have proposed several fraud detection models using ensemble machine learning. In 2018, A Mishra used various ensemble techniques like Gradient boosted tree, Random Forest and stacker classifier with few other classification algorithms to build a machine learning model with the capability of detecting fraud more accurately. The dataset was sampled by performing undersampling and SMOTE as an oversampling technique. They observed that recall and accuracy were improved for all the classifiers with the sampled dataset and gradient boosted ensemble model was able to perform effectively on both actual and sampled dataset [22]. In other research, the ensemble machine learning technique was used with some supervised and unsupervised classifiers for credit card fraud detection. The selected algorithms were applied on a highly imbalanced dataset which consisted of only 492 fraud transactions out of 284,807 total transactions. In this research, they concluded that the unsupervised algorithms can handle the dataset skewness more effectively than the other two applied techniques in case of a highly imbalanced dataset [23]. Similarly, [24] proposed an ensemble learning framework based on the C4.5 algorithm. The original dataset was balanced using the partitioning and clustering approach. In the partitioning step, the dataset was partitioned into training and test set and then the majority class of the training set was randomly divided. Further, clustering was performed on all the majority class blocks. Finally, the nearest neighbor of the centroid of each class was combined with the minority class. As a result, n number of balanced datasets were generated. The C4.5 algorithms were used as base estimators and were trained in parallel on each set of the balanced dataset. The votes of the base estimators were used to build the final classifier. The results observed in the experiment were in the form of improved evaluation metrics: AUC and savings rate.

3   Methodology

This section will describe the process that was followed to implement Supervised, Unsupervised and Deep learning models and to do a hybrid analysis of all the implemented models in detecting fraudulent transactions. To identify fraudulent transactions effectively, the original dataset was processed. In the pre-processing phase, feature extraction and selection were performed on the original dataset. Since the selected dataset was highly imbalanced, we tried to reduce the skewness by using different sampling techniques to balance the original dataset. In our experiment, we have analyzed the performance of Supervised, Unsupervised and Deep learning models in detecting fraudulent transactions and then we have built an ensemble machine learning model using bagging ensemble technique.

3.1 Dataset Information

The dataset for Credit Card Fraud detection is obtained from the Kaggle website which consists of transactions made by credit cards in September 2013 by European cardholders. The dataset consisted of 31 numerical features out of which 28 were named as v1-v28 to protect sensitive data. The remaining 3 columns were represented as Time, Amount and Class. Feature 'Time' contains the seconds elapsed between each transaction and feature 'Amount' is the amount of the transactions made by credit card. Feature 'Class' takes only 2 values: 1 for fraud transaction and 0 for the benign transaction [25].

3.2 Feature Selection

Feature selection is the process wherein relevant features are selected to reduce overfitting, improve accuracy and reduce training time. For feature selection, we used extra tree classifiers – extremely randomized trees, which extracted features on the basis of their scoring using an ensemble learning technique [26].

3.3 Data Balancing

The machine learning algorithm does not learn well when classification categories are not equally distributed. To efficiently train our machine learning classifier balancing plays an important role. Common methods for adjusting the class distribution include under sampling the majority class, oversampling minority class or combination of both the techniques. SMOTE (Synthetic Minority Oversampling Technique) was used to increase the number of instances for our fraud transactions to make our dataset balance. SMOTE implements a statistical technique for increasing the number of fraud cases by generating new instances from existing minority cases that you supply as input. Scaling is performed to bring all features to the same level excluding the values of time and amount as they highly vary. The new balanced dataset consists of a very high number of instances which increases algorithms processing speed. We performed under sampling on our dataset using the Centroid based clustering method. Cluster centroid is an object that under samples majority by replacing the cluster of samples by the cluster centroid of a k-means algorithm. The majority of samples are then completely replaced by the set cluster centroids from K means which provides us with a balanced stable dataset.

3.4 Machine Learning Classifiers

Finally, different machine learning classifiers were used, and results were recorded. All training and testing of models were performed on a 2.20 GHz i7 processor which are discussed in detail below:

3.4.1 Logistic Regression

Logistic regression is also known as the log-linear classifier or maximum-entropy classification. It states that a categorical dependent variable can be predicted from a given set of independent variables. It uses a logistic function to model the possibilities which describe the possible result of a trial. The algorithm behind logistic regression uses a linear decision surface and hence cannot solve nonlinear problems [27].

3.4.2 Decision Tree

A decision tree is a supervised Machine Learning(ML) algorithm which has an inverted tree structure, wherein each internal (non-leaf) node represents feature (predictor variable), the branches between the nodes represent the test outcome, and each leaf (terminal) node represents a class label (response variable). The decision tree classifier makes the decision based on entropy.

Entropy: It measures the uncertainty in the data. Homogeneity of sample data is measured by Decision tree algorithms using entropy. Entropy zero means the sample is fully homogeneous and one means it is equally divided [28].

$$E(S) = \sum_{i=1}^{c} -p_i \log_2 p_i \tag{1}$$

Equation 1 Here pi is the probability of class i, Entropy is computed as the proportion of class i in the set.

3.4.3 Support Vector Machine

Support Vector Machine is also a supervised classifier that can be used for classification. It separates different groups by forming decision boundaries (multi-dimensional space that separates outs classes). We perform classification by finding the hyperplane that differentiates our benign and fraud transactions as shown in the Fig. 1 below:

Fig. 1 Support vector
machine [29]

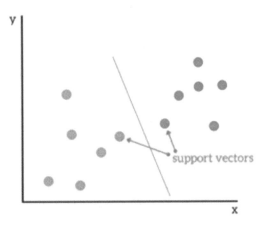

The plane which gives maximum margins to all the categories is labelled as a hyperplane in SVM [30].

3.4.4 K-nearest neighbor

K-nearest neighbors (KNN) is a supervised machine learning algorithm that is used to solve the classification problem. The results of KNN is mainly dependent on 3 factors: the distance metric used to decide the nearest neighbor, the distance rule that is used for the classification from K-nearest neighbor and the number of neighbors considered to classify the new sample.

KNN is largely used as a detection algorithm as it can detect fraudulent transactions with a high rate of performance [18].

3.4.5 K-means

K-means clustering is a popular unsupervised machine learning algorithm used for data classification. It uses a clustering technique to find groups of similar data points aggregated together based on certain similarities. In K-means each cluster is associated with a centroid which allocates every data point to its nearest cluster [31]. It is a distance-based algorithm that aims to minimize the sum of distances between the points and their respective cluster centroid as shown in Fig. 2 below.

3.4.6 Deep Neural Network

Deep neural networks are providing promising results in many fields and are well known for binary classification. Fraud detection is a binary classification problem wherein transaction is analyzed and classified as legitimate or fraudulent.

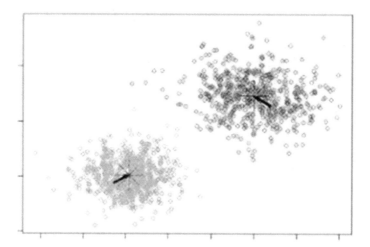

Fig. 2 K-means clustering [32]

Fig. 3 Deep Neural Network [35]

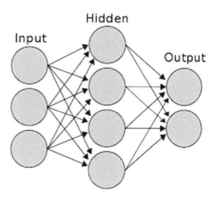

Deep learning using an Artificial neural network (ANN) is a multilayer fully connected neural network which consists of multiple input nodes, where each node is connected to multiple other nodes in the next layer. The model works from left to right (as shown in Fig. 3) in which the weighted sum of its input is passed through a non-linear activation function. Each layer of node trains on a distinct set of features based on the previous layer's output [33, 34].

3.5 Ensemble Classifier

Our primary goal is to reduce the number of false-positive which can be achieved by using an ensemble learning technique. Ensemble learning is based on the principle of combining a group of weak learners that come together to form a strong learner.

As a diverse set of models provide a better result compared to a single model [36]. Most papers have used homogenous base learners (decision trees or random forest) to gain an optimal machine learning algorithm. Our approach is based on heterogeneous learners wherein multiple base learning algorithms are used to give us a strong learner. Wherein several base learners- Decision Tree, Logistic regression, Support Vector Machine and K-nearest neighbor are used to produce one stable predictive optimal model.

3.5.1 Bagging for Supervised Classifier

Bagging is a method of the ensemble learning which groups several machine learning models (base learners) with different samples of the dataset to reduce variance and overfitting in our predictive machine learning model. The original dataset is divided into multiple data set by using sampling with replacement techniques. By using sampling with replacement techniques, the training data is changed for every base learner and in turn, the prediction is different for every base learner. The Scikit-learn function is used to resample the dataset. A combined classifier is produced by taking average prediction of all the base learner classifiers. Predictions are based on using the max voting technique wherein the prediction of each model is taken as a vote. The maximum predictions which we get from the majority of the base models are used as a final prediction to create our hybrid heterogeneous machine learning model [37].

4 Experiment & Results

In this section, we have highlighted the results achieved with different machine learning techniques in performing fraud detection detect fraudulent transactions and will describe the various measures used to evaluate the performance of basic and hybrid machine learning algorithms in fraud transaction detection.

4.1 Evaluation Measures

To measure the performance of the applied machine learning method, the confusion matrix can be used by relating the actual and predicted outcomes of a model: True Positive (TP), True Negative (TN), false-positive (FP) and False Negative (FN) (Table 1). These values can be used to calculate the different metrics in the machine learning algorithm. We have used the four widely used metrics: Accuracy (ACC), True Positive Rate (TPR), False Positive Rate (FPR), ROC and AUC as evaluation metrics [38].

Table 1 Confusion matrix

Actual Class	Predictive	
	True Positive	False Positive
	False Negative	True Negative

Table 2 Observation with supervised classifiers

Model	ACC	TPR	FPR	AUC
Logistic regression	0.75	0.52	0.54	0.98
Decision tree	0.99	0.51	0.52	0.87
Support vector machine	0.77	0.52	0.55	0.96
K-nearest neighbor	0.96	0.85	0.14	0.92

$$\text{Accuracy (ACC)} = \frac{TP + TN}{TP + TN + FP + FN}$$

$$\text{True Positive Rate (TPR)} = \frac{TP}{TP + FN}$$

$$\text{False Positive Rate (FPR)} = \frac{FP}{FP + TN}$$

ROC – is created by plotting a graph of TPR on Y-axis and FPR on X-axis at various thresholds.

AUC – It is measured as an area under the ROC curve. A model with AUC equal to 1 is an ideal model.

4.2 Experiment & Results

We have used the sampled dataset with a total of 60,000 samples: 30,000 fraud and 30,000 normal transactions, which was generated after processing the original dataset. All the models were trained on the sampled dataset and the results were marked based on the selected metrics.

4.2.1 Supervised Model Results

On training our supervised classifier on the sampled dataset, KNN outperformed the other applied supervised classifiers with the highest accuracy of 99.2% (Table 2).

As discussed in Sect. 4.1, the ROC curve is a graph displaying the overall performance of the classification model. The ROC curve for all the supervised classifiers with AUC as the area under the ROC is presented in Figs. 4, 5, 6, and 7. As shown above RCO figures, the AUC observed for all the classifiers are almost

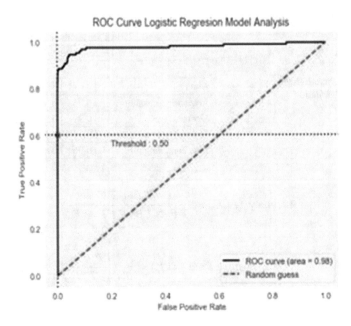

Fig. 4 ROC and AUC – Logistic regression

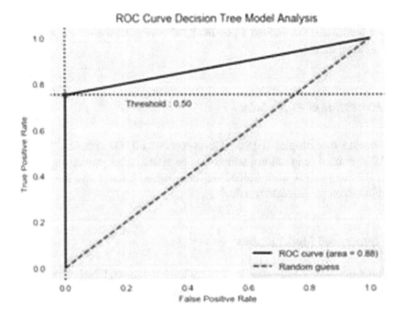

Fig. 5 ROC of Decision Tree

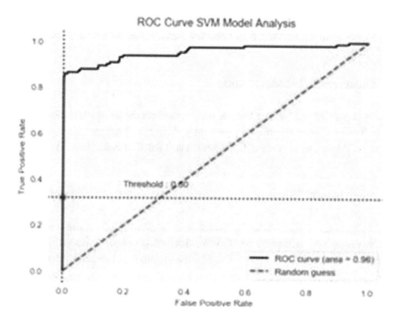

Fig. 6 ROC and AUC of SVM model

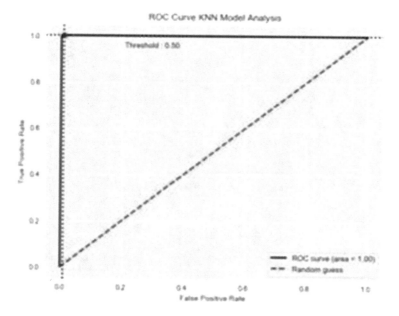

Fig. 7 ROC and AUC of KNN model

close to the ideal value of AUC (AUC = 1). However, KNN came up as the most powerful model in classifying fraud and normal transactions with AUC equal to 1.

4.2.2 Unsupervised Model Results

For the unsupervised models, Kmeans was trained on the sampled dataset. As shown in Table 3, K-means was able to score only 70% accuracy in detecting fraudulent transaction. Also, the lowest AUC was recorded for K-means (Fig. 8).

4.2.3 Deep Neural Network

In deep learning, the deep neural network was implemented on the original dataset and achieved a high accuracy of 99.9% with Deep neural network with a highly imbalanced dataset with a 12 low FPR of 44% (Table 4). However, the AUC metrics was low (Fig. 9).

Table 3 Observation of K-means model

Model	Accuracy	TPR	FPR	AUC
K-means	0.70	0.70	0.27	72

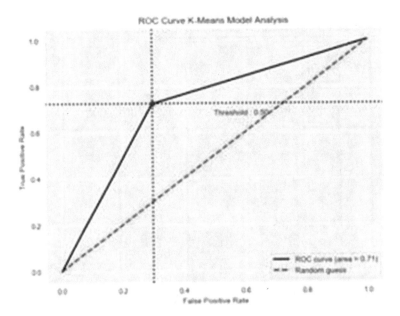

Fig. 8 ROC and AUC for K-means

Table 4 Observation for Deep Neural network

Model	Accuracy	TPR	FPR	AUC
Deep neural networks	0.99	0.99	0.4	0.8

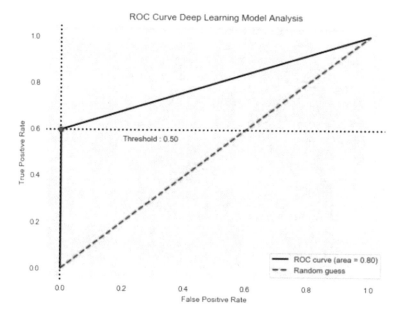

Fig. 9 ROC and AUC for Deep Neural Network

4.2.4 Supervised Ensemble Model – Bagging

The hybrid model built on a supervised classifier outperforms all other machine learning techniques. As shown in Fig. 10. The overall performance of the ensemble model is extremely good with AUC equal to 1 and the accuracy of 99.9% and lowest FPR of 0.01% (Table 5).

5 Conclusion

In this paper, we have performed a comparative analysis of different categories of machine learning: supervised, unsupervised, deep learning and ensemble machine learning in detecting fraud transactions based on their accuracy, TPR, FPR, ROC and AUC evaluation metrics. Conclusion of this paper summarized as follows:

1. Data sampling was done to reduce the skewness of the original imbalanced dataset.
2. The supervised classifiers: KNN, Decision Tree, SVM and Logistic Regression were trained on the sampled dataset and the performance of these classifiers was analyzed based on evaluation metrics.

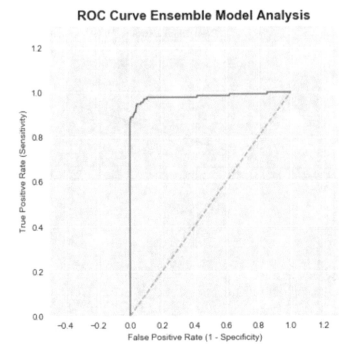

Fig. 10 ROC and AUC for supervised ensemble model

Table 5 Observations for supervised ensemble model

Model	Accuracy	TPR	FPR	AUC
Supervised ensemble-bagging	0.999	0.998	0.015	0.999

3. Implemented ANN in deep learning using the original dataset and observed that it can effectively identify credit card fraud transactions from a highly imbalanced like European Dataset.
4. Implemented an ensemble model by combining the selected supervised classifiers to detect fraud transactions more accurately with minimum false positive rate.

  The future work will be to evaluate the transactions in real-time. We can also implement and analyze the Boosting method of ensemble machine learning with supervised, unsupervised and deep learning.

References

1. B. Peter, Number of credit cards and credit card holders (WalletHub), https://wallethub.com/edu/number-of-credit-cards/25532/. Accessed 16 Sept 2020

2. Credit card fraud statistics. Shift credit card processing, https://shiftprocessing.com/creditcard-fraud-statistics/. Accessed 17 Sept 2020

3. M. Amrollahi, A. Dehghantanha, R.M. Parizi, A survey on application of big data in fin tech banking security and privacy, in *Handbook of Big Data Privacy*, (Springer, Cham, 2020), pp. 319–342

4. A. Yazdinejad, R.M. Parizi, A. Dehghantanha, K.-K.R. Choo, Blockchain-enabled authentication handover with efficient privacy protection in SDN-based 5G networks. IEEE Trans. Netw. Sci. Eng. (2019). https://doi.org/10.1109/TNSE.2019.2937481

5. F. C. A. of Canada, Credit card fraud (aem, 2017, January 10), https://www.canada.ca/en/financialconsumer-agency/services/credit-fraud.html. Accessed 17 Sept 2020

6. A. Yazdinejad, R.M. Parizi, A. Dehghantanha, K.-K.R. Choo, P4-to-blockchain: A secure blockchain-enabled packet parser for software defined networking. Comput. Secur. **88** (2020). https://doi.org/10.1016/j.cose.2019.101629

7. World Health Organization, et al., A. Ligaya, Massive security breach at Capital One exposes data of 6 million Canadians (CTVNews, 2019, July 30), https://www.ctvnews.ca/business/massive-security-breach-at-capital-oneexposes-data-of-6-million-canadians-1.4529639. Accessed 17 Sept 2020. *Osteoarthr. Cartil*

8. R. R. F. DeFilippi, Boosting, bagging, and stacking – Ensemble methods with sklearn and mlens (Medium, 2018, August 4), https://medium.com/@rrfd/boosting-bagging-and-stacking-ensemblemethods-with-sklearn-and-mlens-a455c0c982de. Accessed 10 May 2020

9. A. Yazdinejad, R.M. Parizi, G. Srivastava, A. Dehghantanha, K.-K.R. Choo, Energy efficient decentralized authentication in internet of underwater things using blockchain, in *2019 IEEE Globecom Workshops (GC Wkshps)*, (2019), pp. 1–6

10. A. Yazdinejad, H. HaddadPajouh, A. Dehghantanha, R.M. Parizi, G. Srivastava, M.-Y. Chen, Cryptocurrency malware hunting: A deep recurrent neural network approach. Appl. Soft Comput. J. Elsevier **96**, 106630 (2020)

11. M. Aledhari, R. Razzak, R.M. Parizi, F. Saeed, Federated learning: A survey on enabling technologies, protocols, and applications. IEEE Access **8**, 140699–140725 (2020). https://doi.org/10.1109/ACCESS.2020.3013541

12. A. Yazdinejad, R.M. Parizi, A. Dehghantanha, H. Karimipour, G. Srivastava, M. Aledhari, Enabling drones in the internet of things with decentralized blockchain-based security. IEEE Internet Things J., 1 (2020). https://doi.org/10.1109/jiot.2020.3015382

13. V. Mothukuri, R.M. Parizi, S. Pouriyeh, Y. Huang, A. Dehghantanha, G. Srivastava, A survey on security and privacy of federated learning. Futur. Gener. Comput. Syst. (2020). https://doi.org/10.1016/j.future.2020.10.007

14. A. Yazdinejad, S. Kavei, S.R. Karizno, Increasing the performance of reactive routing protocol using the load balancing and congestion control mechanism in MANET. Comput. Knowl. Eng. **2**(1), 33–42 (2019). https://doi.org/10.22067/cke

15. Q. Chen, G. Srivastava, R.M. Parizi, M. Aloqaily, I. Al Ridhawi, An incentive-aware blockchain-based solution for internet of fake media things. Inf. Process. Manag., 102370 (2020). https://doi.org/10.1016/j.ipm.2020.102370

16. A. Yazdinejad, R.M. Parizi, A. Dehghantanha, G. Srivastava, S. Mohan, A.M. Rababah, Cost optimization of secure routing with untrusted devices in software defined networking. J. Parallel Distrib. Comput. **143**, 36–46 (2020)

17. A. Yazdinejad, R.M. Parizi, A. Bohlooli, A. Dehghantanha, K.-K.R. Choo, A high-performance framework for a network programmable packet processor using P4 and FPGA. J. Netw. Comput. Appl. **156**, 102564 (2020)

18. J.O. Awoyemi, A.O. Adetunmbi, S.A. Oluwadare, Credit card fraud detection using machine learning techniques: A comparative analysis, in *2017 International Conference on Computing Networking and Informatics (ICCNI)*, (2017), pp. 1–9

19. S. Dhankhad, E. Mohammed, B. Far, Supervised machine learning algorithms for credit card fraudulent transaction detection: A comparative study, in *2018 IEEE International Conference on Information Reuse and Integration (IRI)*, (2018), pp. 122–125

20. A. Thennakoon, C. Bhagyani, S. Premadasa, S. Mihiranga, N. Kuruwitaarachchi, Real-time credit card fraud detection using machine learning, in *2019 9th International Conference on Cloud Computing, Data Science Engineering (Confluence)*, (2019, January). https://doi.org/10.1109/CONFLUENCE.2019.8776942

21. T. Choudhury, G. Dangi, T.P. Singh, A. Chauhan, A. Aggarwal, An efficient way to detect credit card fraud using machine learning methodologies, in *2018 Second International Conference on Green Computing and Internet of Things, ICGCIoT 2018*, (2018). https://doi.org/10.1109/ICGCIoT.2018.8753077

22. A. Mishra, C. Ghorpade, Credit card fraud detection on the skewed data using various classification and ensemble techniques, in *2018 IEEE International Students' Conference on Electrical, Electronics and Computer Science (SCEECS)*, (2018), pp. 1–5

23. S. Mittal, S. Tyagi, Performance evaluation of machine learning algorithms for credit card fraud detection, in *2019 9th International Conference on Cloud Computing, Data Science & Engineering (Confluence)*, (2019), pp. 320–324

24. H. Wang, P. Zhu, X. Zou, S. Qin, An ensemble learning framework for credit card fraud detection based on training set partitioning and clustering, in *2018 IEEE SmartWorld, Ubiquitous Intelligence & Computing, Advanced & Trusted Computing, Scalable Computing & Communications, Cloud & Big Data Computing, Internet of People and Smart City Innovation (SmartWorld/SCALCOM/UIC/ATC/CBDCom/IOP/SCI)*, (2018), pp. 94–98

25. Credit card fraud detection, https://kaggle.com/mlg-ulb/creditcardfraud. Accessed 16 Sept 2020

26. 3.2.4.3.3. sklearn.ensemble.ExtraTreesClassifier – scikit-learn 0.22.2 documentation, https://scikit-learn.org/stable/modules/generated/sklearn.ensemble.ExtraTreesClassifier.html. Accessed 10 May 2020

27. Logistic regression in python | Python for data science (Edureka, 2019, April 16), https://www.edureka.co/blog/logistic-regression-in-python/. Accessed 17 Sept 2020

28. Decision tree classification in python (DataCamp Community, 2018, December 28), https://www.datacamp.com/community/tutorials/decision-tree-classification-python. Accessed 17 Sept 2020

29. Support vector machines: A simple explanation (KDnuggets), https://www.kdnuggets.com/support-vector-machines-a-simple-explanation.html/. Accessed 17 Sept 2020

30. A. Yadav, Support Vector Machines (SVM). Introduction: All you need to know . . . | by Ajay Yadav | Towards Data Science

31. K means clustering | K means clustering algorithm in python (Analytics Vidhya, 2019, August 19), https://www.analyticsvidhya.com/blog/2019/08/comprehensive-guide-k-means-clustering/. Accessed 17 Sept 2020

32. P. Jeffcock, K-means clustering in machine learning, simplified, https://blogs.oracle.com/bigdata/k-means-clustering-machine-learning. Accessed 17 Sept 2020

33. World Health Organization, et al., A. Dertat, Applied deep learning – Part 1: Artificial neural networks (Medium, 2017, Ocotber 9), https://towardsdatascience.com/applied-deep-learning-part-1-artificial-neural-networks-d7834f67a4f6. Accessed 17 Sept 2020. *Osteoarthr. Cartil*

34. A beginner's guide to neural networks and deep learning – Pathmind, https://pathmind.com/wiki/neural-network. Accessed 26 Jan 2020

35. ANN algorithm | How artificial neural network works (Analytics Vidhya, 2014, October 20), https://www.analyticsvidhya.com/blog/2014/10/ann-work-simplified/. Accessed 17 Sept 2020

36. J. Shubham, Ensemble learning – Bagging and boosting (Medium, 2018, July 6), https://becominghuman.ai/ensemble-learning-bagging-and-boosting-d20f38be9b1e. Accessed 17 Sept 2020

37. Ensemble learning in python (DataCamp Community, 2018, September 6), https://www.datacamp.com/community/tutorials/ensemble-learning-python. Accessed 17 Sep 2020

38. M. Sokolova, N. Japkowicz, S. Szpakowicz, Beyond accuracy, F-score and ROC: A family of discriminant measures for performance evaluation, in *Australasian Joint Conference on Artificial Intelligence*, (2006), pp. 1015–1021

Mapping CKC Model Through NLP Modelling for APT Groups Reports

Aaruni Upadhyay, Samira Eisaloo Gharghasheh, and Sanaz Nakhodchi

1 Introduction

Many business establishments still rely on outdated security measures to counter security threats. The effectiveness of such traditional approaches has diminished over time and that becomes even more evident when organizations face advanced cybercriminals like APT [1–3] actors. APT attacks are carried by extremely accomplished (possibly state-sponsored) cybercriminal teams who have doubtless unlimited time and resources [4].

APTs are known to compromising again and again people, companies and governments, by using various techniques and ways to realize their targets [2, 5]. "APT became famous following a New York Times exposé detailing a month's long attack campaign in which a Chinese military unit now known as "APT 1" thoroughly penetrated the media organization's networks with a series of spear-phishing emails and a deluge of customized malware samples." APT can appear in two ways: APT as a factor or individuals. From one side, APT points to an extremely accurate cyberattack. On the opposite side, the APT can even be the teams, typically "state-sponsored or well-funded" in different methods, which they are able to fire well-aimed attacks [3].

According to the last cyber-attacks, research in this field has been increased tremendously [6–11]. These attacks have widespread destructive effects on their targets [1, 12–14]. CKC is a model to make transparency for incident response team and security analysts to have a better understanding of the incidents [4].

CKC has seven phases which includes Reconnaissance, Weaponization, Delivery, Exploitation, Installation, Command & Control(C2) and Actions on Objectives

A. Upadhyay · S. E. Gharghasheh · S. Nakhodchi (✉)
School of Computer Science, Computer Science, University of Guelph, Guelph, ON, Canada
e-mail: aupadhya@uoguelph.ca; samira@cybersciencelab.org; sanaz@cybersciencelab.org

© Springer Nature Switzerland AG 2021
K.-K. R. Choo, A. Dehghantanha (eds.), *Handbook of Big Data Analytics and Forensics*, https://doi.org/10.1007/978-3-030-74753-4_16

(AOO). A typical cyberattack begins with information gathering on the target as its first step (reconnaissance). The details of enemy vulnerabilities may be used to develop a malware (weaponization) and can also be used to plan the most undetectable way of infecting victim machines (delivery). Thereafter the attacker wants to carry out the attack (exploitation) and gain persistence (installation). Once these stages are successfully executed, the attacker is able to control its attack remotely (C2) and is able to carry out its exploitation at will (AOO). On the other hand, during the past years cybercommunity has witnessed a significant increase of leveraging machine learning methods to provide successful approaches ranging from malware detection [15–19] and malware hunting [20–24] to privacy [25, 26] and intelligent cyber-analytics [27].

In this research, we present an automatic way to process the APT reports using NLP techniques to identify and map the different CKC stages employed in the attack. We use the Latent Semantic Indexing (LSI) and Latent Dirichlet Allocation (LDA) to automatically extract information related to CKC stages from the APT reports. We then compare our results with a recent survey that classified APT groups into different CKC stages manually. In this research we present LSI and LDA as a faster alternative and discuss the ways to achieve better accuracy in results.

This paper organized as follows. Section 2 presents the previous work on this field. Then Sect. 3 provides the methodology. The results are discussed and evaluated in Sect. 4. Finally, this paper is concluded in Sect. 5.

2   Related Works

NLP models usually are using for keyword extraction, topic modeling and so on. Since, security reports are unstructured text, NLP models are useful for data extraction from technical reports. For example, [28] proposed SECCMiner which is an information retrieval system to achieve useful information about APTs from their unstructured reports. They used NLP and information retrieval methods to extract information such as the techniques and tactics of attacks. They collected 445 APT reports with about 1.9 million words between 2008 and 2017. [29] used Latent Semantic Analysis for capturing keywords from cyber security weblogs with focusing on certain topics. They focused on searching blog for topics such as cybercrime, cyber terrorism and cyber security threats. In addition, their method also tracked prominent "conversation and topics in the blogsphere." "a subset of the Nielson BuzzMetrics weblog data corpus" used for the dataset and their experiments. The keyword detection and weblog search were boost for information retrieval. The generic detecting method proposed in [30] which used unsupervised algorithm to learn the context of proxy server log. The Paragraph Vector learnt feature representation with fixed-length from variable-length pieces of texts. They implemented algorithm on MWS datasets with D3M and BOS for analyzing timeline. They achieved 0.99 and 0.98 for f-measure of un-known drive-by-download attacks and C2 traffic, respectively.

Automated model for identifying IOCs from reports based on a neural sequence labeling used in [31]. It was an end-to-end sequence labeling for IOCs identification. The dataset included 687 cybersecurity articles related to APT reports from 2008 to 2018. Moreover, 90.4% and 87.2% were precision and recall of their achievement which were better than previous sequence labeling models. [32] used LDA model for vulnerability reports on the Common Vulnerability and Exposures (CVE) database. They implemented their models on description section of reports for finding vulnerability types and novel trends. 39,393 CVE were considered.

Based on formation above, most of the researches have been utilized NLP models for extraction information with different perspective although finding the attack pattern with NLP approach is neglected. Thus, in this research, we used NLP models such as LSI and LDA for mapping the CKC model to APT reports.

3 Methodology

In this section, the method for extracting CKC model from unstructured APT reports is elaborated. In addition, the used dataset, collection method and pre-processing steps for implementation are described.

3.1 Dataset

Our dataset consists of APT reports collected across several sources such as Fireeye, McAfee, Kaspersky etc. We used [33] as our dataset with a total of 108 files for a total of 19 APT groups.

3.2 Data Collection

Our base paper had raw reports related to APT groups up to APT33 and even more reports have since been released. We downloaded the new reports which were APT 34, APT 37, APT 38, APT 40, APT41 and converted them to the .txt file and add to the dataset. In addition, based on Fireeye and Mitre, there were some additional reports that was added such as APT5, APT10, APT17, etc. The primary challenge was that the reports were distributed across several text files. For example, there were different 10 reports for APT1. We combined all similar APT files into a single file so that we have one file for each APT. For the final step of consolidation, we merged all the 19 APT files into a single TXT file with each line of the file representing an APT group.

3.3   Pre-processing

LSI and LDA work on a bad of word model which involves breaking down the text into individual words. The order of words does not matter. Also, the stop words like "an", "the", "is" etc. which are very common in English language are ignored. The next step is to ignore any words that appear only once in the document as they do not help in semantic search. The next step involves building up a dictionary that assigns a unique integer key to each word in the document. This index is used in the corpus that is basically a tuple with the index as first value and frequency of that word in its document as the second. For example, (8, 12) tells that word represented by index 8 appears 12 times in its corresponding document.

3.4   Implementation

We selected LSI and LDA which are popular models in topic modeling. LSI is one of NLP technique for semantically analyzing relationships between "a set of documents and the terms they contain by producing a set of concepts related to the documents and terms." Similar words in terms of meaning occur in similar pieces of text is one of the assumptions of this model [34] . LDA is another model for topic modeling in NLP which is used to classify a document to a specific topic. This model assumes that a document mixes with several topics in it and the topics relate with the words [35]. The main difference of these two models are their calculation, LSI uses Singular Value Decomposition (SVD) matrix on the terms of document however LDA based on probabilistic model.

For our search we created a text file with the APT features based on [12] which can be seen in the Fig. 1.

We use GenSim [36] library for our implementation of LSI and LDA. The first step involves building the LSI model object by passing it the dictionary and the corpus object. Next is the step of performing a document similarity search on this object by using the Gensim's similarity module. This is done into three different steps, first we take the input search stream and break it down into individual words. Next, we convert it to the LSI space by passing it to the LSI model object we created earlier. Next, we perform a similarity query against the corpus and then sort our result in the decreasing order of similarity score.

We now use LSI and LDA to search for CKC stages inside the APT files and create a map. For this we finalize a list of search queries that are closest to the CKC stages.

These queries are listed in the Table 1. The results for each of search query were recorded and used to build the below table that has a "X" for each APT group that appeared in a search query with an accuracy >0.8. The experiment was running for all stages using both LSI and LDA, the results of which are shown in Tables 2 and 3 below.

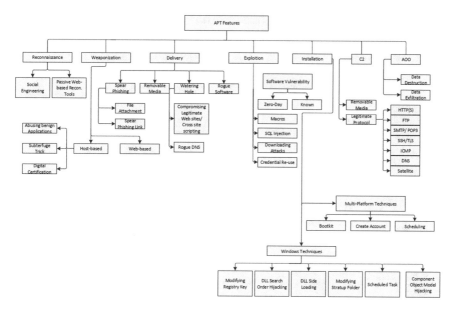

Fig. 1 APT features based on CKC model [14]

The below grid shows the CKC stages across which the different APT groups falls according to LSI/LDA models. However, it does not show the order of relevance of an APT group in the CKC stages. For example, looking at above table, we cannot tell which APT group is most relevant under "social engineering" stage of CKC model. To address this question, we built the heatmaps for both the LSI and LDA models as shown below in Figs. 2 and 3 below.

4 Evaluation and Findings

In this section we present the evaluation of our proposed methodology. In particular, our semantic search based on LSI and LDA algorithms is evaluated for its effectiveness in mapping APT documents to the various CKC stages. We compare our results with [12] and compute the evaluation parameters Precision and Recall.

The following is the details of how we calculated the Precision and Recall parameters:

True Positive If an APT group is mapped to a CKC stage by both our method and correspondingly in the base paper result.

False Positive If an APT group is mapped to a CKC stage by our method but not in the base paper result.

Table 1 Mapping between CKC stage and search term used

CKC stage	Search text
Social engineering	Social engineering
Passive web-based tools	Reconnaissance web tool
Abusing benign software applications vulnerabilities	Software vulnerability
Subterfuge techniques	Subterfuge technique
Legitimate digital certification	Stolen digital certificates
Website equipping	Compromised website
Malicious file attachment	Malicious file attachment
Spear phishing link	Spear phishing link
Compromised legitimate websites	Watering hole website
Rogue DNS	Rogue dns
Replicate through removable media	Replicate through removable media
Rogue software	Rogue software
0-day exploits	0-day exploit
Known exploits	Known exploits
SQL injection	SQL injection
Macros	Exploitation macros
Drive by download	Drive by download exploit
Credential reuse	Credential reuse exploit
Modifying registry key	Installation modifying registry keys
DLL search order hijacking	DLL search order hijacking
DLL side loading	Dll side loading attack
Startup folder	Installation startup folder
Scheduling task	Installation scheduling task
COM hijacking	Com hijacking
Bootkit	Install bootkit
Create account with valid credential	Create account with valid credential
Local job scheduling	Linux macos job scheduling
HTTP(s)	c2 http https tunneling
FTP	c2 ftp tunneling
SMTP/POP3	c2 smtp pop3 tunneling
SSH/TLS	c2 ssh tls tunneling
ICMP	c2 icmp tunneling
DNS	c2 dns tunneling
Satellite	c2 satellite tunneling
C2's using removable media	c2 removable media
Data exfiltration	Data exfiltration objective
Data destruction	Data destruction objective

False Negative If an APT group is not mapped to a CKC stage by our method but is mapped correspondingly in the base paper.

True Negative If an APT group is not mapped to a CKC stage and also not mapped correspondingly in the base paper.

Table 2 Our results using LSI model

APT Groups	Reconnaissance		Weaponization				Delivery					Exploitation							Installation									Command and Control (C2)							Actions on objectives		
			Host-based			Network-based	Email/local network spear phishing			Watering hole			Software vulnerabilities						Windows-platform techniques						Multi-platform techniques			Tunneling over network protocols									
	Social Engineering	Passive web-based tools	Abusing benign software vulnerabilities	Subterfuge applications	Legitimate digital certification	Website spoofing	Malicious File attachment	Spear phishing link	Compromised legitimate websites	Rogue DNS	Replicate through removable media	Rogue Software	0-day Exploits	Known Exploits	SQL injection	Macros	Drive by download	Credential reuse	Modifying Registry key	DLL Search Order Hijacking	DLL side loading	Startup folder	Scheduling task	COM Hijacking	Rootkit	Create account with valid credential	Local Job Scheduling	HTTP(s)	FTP	SMTP/POP3	SSH/TLS	ICMP	DNS	Satellite	C2 using removable media	Data exfiltration	Data destruction
APT1	x	x	x			x	x	x	x		x		x	x	x	x	x		x	x						x	x	x		x						x	x
APT3			x			x	x	x	x		x		x	x		x	x		x	x						x	x	x						x	x	x	x
APT5	x	x	x			x	x	x	x		x		x	x		x	x		x	x	x	x				x	x		x		x	x		x	x	x	x
APT10	x	x	x		x	x	x	x	x		x		x	x			x		x	x	x	x				x	x		x		x				x	x	x
APT12	x		x			x	x	x	x		x		x	x			x		x	x			x			x	x								x	x	x
APT16			x		x	x	x		x		x		x	x		x	x		x	x	x	x				x	x		x					x	x	x	x
APT17			x	x			x		x		x		x	x			x		x	x						x	x		x		x	x		x	x	x	x
APT18							x		x		x		x	x			x		x								x									x	x
APT28							x	x	x		x		x	x		x	x	x	x								x									x	
APT29			x	x			x	x	x		x		x	x			x	x	x	x						x	x									x	
APT30	x		x				x	x	x		x		x	x			x	x	x	x						x	x			x		x		x	x	x	x
APT32	x	x					x	x	x		x		x	x		x	x		x	x			x			x	x	x	x	x	x	x		x	x	x	x
APT33	x	x					x	x	x	x	x		x	x		x	x		x	x		x	x			x	x	x	x	x	x	x		x	x	x	x
APT34	x	x		x		x	x	x	x		x	x	x	x	x	x	x		x	x	x		x			x	x		x	x	x	x		x	x	x	x
APT37	x	x					x	x	x	x	x		x	x		x	x		x	x	x					x	x	x	x	x	x	x	x	x	x	x	x
APT38	x	x					x	x	x		x		x	x		x	x		x	x						x	x	x	x	x	x			x	x	x	x
APT39	x	x					x	x	x		x		x	x		x	x		x	x						x	x	x	x	x	x			x	x	x	x
APT40	x	x				x	x	x	x		x		x	x		x	x		x	x						x	x		x	x	x			x	x	x	x
APT41	x					x	x	x	x		x		x	x	x	x	x		x	x	x	x	x	x		x	x	x	x	x	x	x	x	x	x	x	x

Table 3 Our results using LDA model

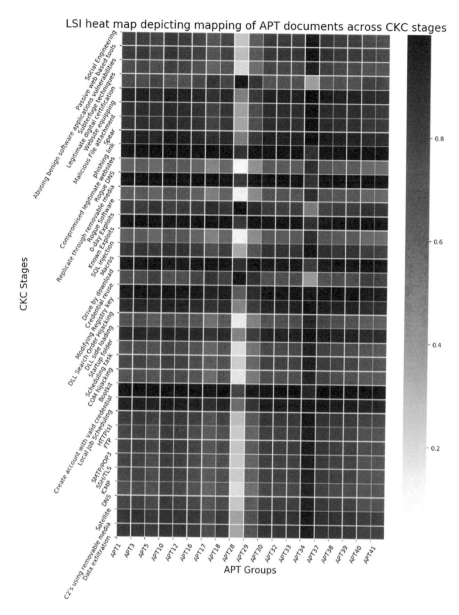

Fig. 2 LSI heat map

The base paper has only 8 APT groups in common to us and we calculated the above evaluation parameters for these APT groups for both LSI and LDA methods. Table 4 presents our evaluation results.

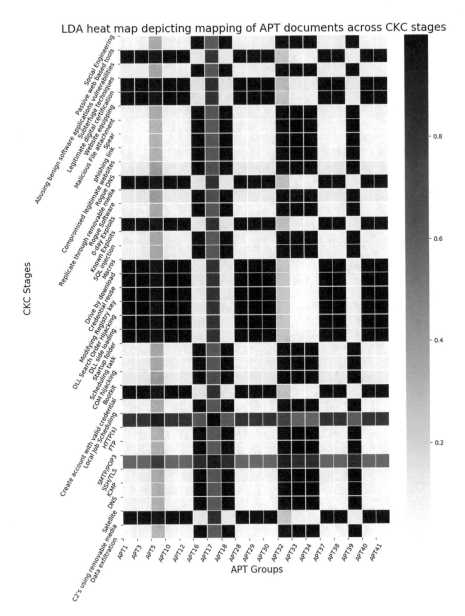

Fig. 3 LDA heat map

We then calculated Precision and Recall using the below formulae to calculate our accuracy:

Precision = *True Positive / (True Positive + False Positive)*
Recall = *True Positive / (True Positive + False Negative)*

Table 4 LSI/LDA evaluation for 8 APT groups

	LSI				LDA			
	True positives	False positives	False negatives	True negatives	True positives	False positives	False negatives	True negatives
APT1	7	13	5	12	6	7	6	18
APT3	13	10	5	9	7	6	11	13
APT12	8	15	2	12	5	8	5	19
APT16	5	13	3	16	5	19	3	10
APT17	9	5	7	16	7	7	9	14
APT28	7	0	15	15	9	4	13	11
APT29	9	5	10	13	8	5	11	13
APT30	9	14	2	12	7	6	4	20
Sum	67	75	49	105	54	62	62	118

Table 5 Comparison
between LSI and LDA

	LSI	LDA
Precision	47%	47%
Recall	58%	47%

Precision helps us tell "what proportion of positive identifications was actually correct" while Recall helps us answer "what proportion of actual positives was identified correctly"?

Table 5 shows the comparison of Precision and Recall for the two models. Our Precision and Recall values were lower than our expectation and we list the below factors that might have had an affect:

1. We have incorporated new APT reports since the base paper was written and as such there will be new CKC stages under which APT groups will fall in. This may have contributed to an increase in False Positives and hence bringing down the Precision score of our result.
2. Selection of Keywords: We attempted to keep our search queries very close to the CKC stage categories developed by the author(s) of the base paper. Perhaps further refining our query to each CKC stage to would yield better result. For example, searching for "phishing" instead of "social engineering" provides 18 APT groups instead of 10.
3. Limited input data: Our sources for APT reports were limited to the APT reports available freely on the internet. We expect that the addition of more input data from proprietary reports will increase the accuracy of our model.
4. We have only 8/19 APT groups in common with the base paper for comparison purpose and as such a full comparison would not be possible.

5   Conclusion

APT reports track the Tactics, Techniques, and Procedures (TTPs) employed by the APT actors in their mission. These reports frequently come in form of unstructured text that must be parsed manually for information. Anyone looking to identify various CKC stages within such a report would need to invest considerable time and energy to do so. Our semantic search method using LSI/LDA for finding CKC stages within an APT report provides a fast alternative to manual labor. Our approach for preprocessing APT reports and running semantic search can be improved upon to build an automatic way of CKC stage classification. For future work, other semantic search models may be applied along with using much bigger set of input data.

References

1. T. Dargahi, A. Dehghantanha, P.N. Bahrami, M. Conti, G. Bianchi, L. Benedetto, A cyber-kill-chain based taxonomy of crypto-ransomware features. J. Comput. Virol. Hacking Tech. **15**(4), 277–305 (2019)
2. S. Grooby, T. Dargahi, A. Dehghantanha, Protecting IoT and ICS platforms against advanced persistent threat actors: Analysis of APT1, silent Chollima and molerats, in *Handbook of Big Data and IoT Security*, (Springer, Cham, 2019), pp. 225–255
3. H. Haddadpajouh, A. Azmoodeh, A. Dehghantanha, R.M. Parizi, MVFCC: A multi-view fuzzy consensus clustering model for malware threat attribution. IEEE Access **8**, 139188–139198 (2020)
4. H. Mwiki, T. Dargahi, A. Dehghantanha, K.-K.R. Choo, Analysis and triage of advanced hacking groups targeting western countries critical national infrastructure: APT28, RED October, and Regin, in *Critical Infrastructure Security and Resilience*, (Springer, Cham, 2019), pp. 221–244
5. A. Yazdinejad, R. M. Parizi, A. Dehghantanha, K.-K. R. Choo, Blockchain-enabled authentication handover with efficient privacy protection in SDN-based 5G networks IEEE Trans. Netw. Sci. Eng. **8**(2), 1120–1132 (1 April–June 2021). https://doi.org/10.1109/TNSE.2019.2937481
6. A. Yazdinejad, H. HaddadPajouh, A. Dehghantanha, R.M. Parizi, G. Srivastava, M.-Y. Chen, Cryptocurrency malware hunting: A deep recurrent neural network approach. Appl. Soft Comput. **96**, 106630 (2020 Nov 1). Elsevier.
7. M. Aledhari, R. Razzak, R.M. Parizi, F. Saeed, Federated learning: A survey on enabling technologies, protocols, and applications. IEEE Access **8**, 140699–140725 (2020).
8. A. Yazdinejad, R.M. Parizi, A. Dehghantanha, H. Karimipour, G. Srivastava, M. Aledhari, Enabling drones in the internet of things with decentralized Blockchain-based security. IEEE Internet Things J., 1 (IEEE, 2020). https://doi.org/10.1109/jiot.2020.3015382
9. V. Mothukuri, R.M. Parizi, S. Pouriyeh, Y. Huang, A. Dehghantanha, G. Srivastava, A survey on security and privacy of federated learning. Futur. Gener. Comput. Syst. **115**, 619 (2020)
10. A. Yazdinejad, A. Bohlooli, K. Jamshidi, Performance improvement and hardware implementation of open flow switch using FPGA, in *2019 5th Conference on Knowledge Based Engineering and Innovation (KBEI)*, (2019), pp. 515–520
11. A. Yazdinejad, S. Kavei, S. Razaghi Karizno, Increasing the performance of reactive routing protocol using the load balancing and congestion control mechanism in MANET. Comput. Knowl. Eng. **2**(1), 33–42 (2019). https://doi.org/10.22067/cke
12. P.N. Bahrami, A. Dehghantanha, T. Dargahi, R.M. Parizi, K.-K.R. Choo, H.H.S. Javadi, Cyber kill chain-based taxonomy of advanced persistent threat actors: Analogy of tactics, techniques, and procedures. J. Inf. Process. Syst. **15**(4), 865–889 (2019)
13. R. HosseiniNejad, H. HaddadPajouh, A. Dehghantanha, R.M. Parizi, A cyber kill chain based analysis of remote access trojans, in *Handbook of Big Data and Iot Security*, (Springer, Cham, 2019), pp. 273–299
14. D. Kiwia, A. Dehghantanha, K.-K.R. Choo, J. Slaughter, A cyber kill chain based taxonomy of banking Trojans for evolutionary computational intelligence. J. Comput. Sci. **27**, 394–409 (2018)
15. A. Azmoodeh, A. Dehghantanha, M. Conti, K.-K.R. Choo, Detecting crypto-ransomware in IoT networks based on energy consumption footprint. J. Ambient. Intell. Humaniz. Comput. **9**(4), 1141–1152 (2018)
16. H. HaddadPajouh, A. Dehghantanha, R. Khayami, K.-K.R. Choo, A deep recurrent neural network based approach for internet of things malware threat hunting. Futur. Gener. Comput. Syst. **85**, 88–96 (2018). https://doi.org/10.1016/j.future.2018.03.007
17. A. Azmoodeh, A. Dehghantanha, K.-K.R. Choo, Robust malware detection for internet of (battlefield) things devices using deep eigenspace learning. IEEE Trans. Sustain. Comput. **4**(1), 88–95 (2018)

18. H. Darabian et al., A multiview learning method for malware threat hunting: Windows, IoT and android as case studies. World Wide Web 23(2), 1241–1260 (2020)
19. M. Saharkhizan, A. Azmoodeh, A. Dehghantanha, K.-K.R. Choo, R.M. Parizi, An ensemble of deep recurrent neural networks for detecting IoT cyber attacks using network traffic. IEEE Internet Things J. 7(9), 8852–8859 (2020). https://doi.org/10.1109/jiot.2020.2996425
20. H.H. Pajouh, R. Javidan, R. Khayami, D. Ali, K.-K.R. Choo, A two-layer dimension reduction and two-tier classification model for anomaly-based intrusion detection in IoT backbone networks. IEEE Trans. Emerg. Top. Comput. 7, 314 (2016)
21. M. Saharkhizan, A. Azmoodeh, H. HaddadPajouh, A. Dehghantanha, R.M. Parizi, G. Srivastava, A hybrid deep generative local metric learning method for intrusion detection, in Handbook of Big Data Privacy, (Springer, Cham, 2020), pp. 343–357. https://doi.org/10.1007/978-3-030-38557-6_16
22. A.N. Jahromi et al., An improved two-hidden-layer extreme learning machine for malware hunting. Comput. Secur. 89, 101655 (2020)
23. A. Azmoodeh, A. Dehghantanha, R.M. Parizi, S. Hashemi, B. Gharabaghi, G. Srivastava, Active spectral botnet detection based on eigenvalue weighting, in Handbook of Big Data Privacy, (Springer, Cham, 2020), pp. 385–397. https://doi.org/10.1007/978-3-030-38557-6_19
24. S. Homayoun, A. Dehghantanha, M. Ahmadzadeh, S. Hashemi, R. Khayami, Know abnormal, find evil: Frequent pattern mining for ransomware threat hunting and intelligence. IEEE Trans. Emerg. Top. Comput. 8, 341 (2017)
25. K.-K.R.C.A. Dehghantanha, Eda, Handbook of Big Data Privacy (Springer, Cham, 2020)
26. A. Ekramifard, H. Amintoosi, A.H. Seno, A. Dehghantanha, R.M. Parizi, A systematic literature review of integration of Blockchain and artificial intelligence, in Advances in Information Security, (Springer, Cham, 2020), pp. 147–160. https://doi.org/10.1007/978-3-030-38181-3_8
27. M. Conti, T. Dargahi, A. Dehghantanha, Cyber threat intelligence: Challenges and opportunities, in Advances in Information Security, (Springer, Cham, 2018), pp. 1–6. https://doi.org/10.1007/978-3-319-73951-9_1
28. A. Niakanlahiji, J. Wei, B.T. Chu, A natural language processing based trend analysis of advanced persistent threat techniques, in Proceedings – 2018 IEEE International Conference on Big Data 2018, (2019 January), pp. 2995–3000. https://doi.org/10.1109/BigData.2018.8622255
29. F.S. Tsai, K.L. Chan, Detecting cyber security threats in weblogs using probabilistic models, in Pacific-Asia Workshop on Intelligence and Security Informatics, (Springer, Berlin/Heidelberg, 2007), pp. 46–57
30. M. Mimura, H. Tanaka, Heavy log reader: learning the context of cyber attacks automatically with paragraph vector, in International Conference on Information Systems Security, (Springer, Cham, 2017), pp. 146–163
31. S. Zhou, Z. Long, L. Tan, H. Guo, Automatic identification of indicators of compromise using neural-based sequence labelling, arXiv Prepr. arXiv1810.10156 (2018)
32. S. Neuhaus, T. Zimmermann, Security trend analysis with cve topic models. in 2010 IEEE 21st International Symposium on Software Reliability Engineering, (IEEE, 2010), pp. 111–120
33. U. Noor, Z. Anwar, T. Amjad, K.-K.R. Choo, A machine learning-based FinTech cyber threat attribution framework using high-level indicators of compromise. Futur. Gener. Comput. Syst. 96, 227–242 (2019)
34. S.T. Dumais, A graph analytical approach for topic detection. Annu. Rev. Inf. Sci. Technol. 38(188) (2005)
35. P. Dwivedi, NLP: Extracting the Main Topics from Your Dataset Using LDA in Minutes (2018). https://towardsdatascience.com/nlp-extractingthe-main-topics-from-your-dataset-using-lda-in-minutes-21486f5aa925. Accessed 30 Nov 2019
36. R. Rehurek, Gensim: Documentation (2019). https://radimrehurek.com/gensim/auto_examples/index.html. Accessed 01 Dec 2019

Ransomware Threat Detection: A Deep Learning Approach

Kassidy Marsh and Hamed Haddadpajouh

1 Introduction

Ransomware is a special type of computer malware which is becoming more and more prevalent in the world of cybercrime [1]. It is estimated that in 2016 alone, more than $1 billion was paid towards ransoms which were associated with ransomware [2]. Ransomware works by preventing access to crucial proprietary data, and only releases access to the data upon payment of a ransom [3]. There are two main types of ransomware: Crypto ransomware and Locker ransomware. Locker ransomware acts by locking a user out of a device until the ransom is paid; the actual data on the device is untouched, and as a result the data can potentially be recovered by removing the ransomware from the device or by transferring the hard drive to a non-infected device [4]. Due to the shortcomings of Locker ransomware (in the eyes of the attacker, that is), Crypto ransomware is much more common [5]. Crypto ransomware involves the encryption of all files on a device, preventing their recovery without a decryption key (which is received upon payment of ransom) [6].

To detect the presence of ransomware, a wide variety of machine learning techniques have been employed over the years [2]. These techniques make use of the frequent patterns in operation code that commonly appear in ransomware applications [7]. Machine learning models can be trained to recognize patterns that are common in ransomware files, and even distinguish between the different families of ransomware. Several papers have been published with high success rates for identifying ransomware files using machine learning [8–10].

K. Marsh (✉)
School of Computer Science, University of Guelph, Guelph, ON, Canada
e-mail: kmarsh08@uoguelph.ca

H. Haddadpajouh
Cyber Science Lab, University of Guelph, Guelph, ON, Canada
e-mail: hamed@cybersciencelab.org

© Springer Nature Switzerland AG 2021
K.-K. R. Choo, A. Dehghantanha (eds.), *Handbook of Big Data Analytics and Forensics*, https://doi.org/10.1007/978-3-030-74753-4_17

253

In this paper, we make use of a dataset consisting of ransomware samples that were previously collected by Homayoun et al. [11]. We extract frequently occurring patterns from these samples and evaluate the performance of different machine learning models at recognizing these patterns. After being trained on the frequently occurring patterns, the objective is that the machine learning models will classify each ransomware sample into one of six possible ransomware families: Cerber, CryptoWall, CTB-Locker, Locky, Sage, TeslaCrypt. We make use of five different machine learning algorithms: K-Nearest Neighbours (KNN), Convolutional Neural Network (CNN), Logistic Regression (LR), Random Forest (RF), and Decision Tree (DT). CNNs are a form of deep learning and therefore have significantly higher computational complexity compared to the other algorithms used here. Performance of algorithms is evaluated using common metrics such as True Positive Rate (TPR), False Positive Rate (FPR), accuracy, Receiver Operating Characteristics (ROC) curves, and the area under the ROC curves (AUROC) [12]. Descriptions of each of these metrics can be found in Sects. 4.1 (TPR/FPR/accuracy) and 4.2.7 (ROC and AUROC).

This work seeks to compare the performance of these five machine learning algorithms when classifying ransomware samples into their respective families. It was predicted that our CNN model will outperform the other models, due to the increased power of CNNs (it is the only deep learning model used in this paper). However, the performance of the other algorithms is still of interest due to the reduced complexity of these algorithms; compared to CNNs, the non-deep learning models are extremely fast, and speed is often important in cases of ransomware detection. Potential contributions from this paper will be a continuation of the previous work by Homayoun et al. [11], where several machine learning models were used to classify the same dataset of ransomware samples.

Section 2 of this paper contains a literature review on similar papers from recent years. Section 3 details the methodology used in this work, Sect. 4 contains the results of our experiments, and in Sect. 5 we draw conclusions and suggest future works. Finally, references are in last section.

2 Related Works

Ransomware is a type of malware that has persisted in the cybersecurity world since the 1980s [13]. The changing nature of ransomware attacks and the introduction of new families of ransomware has caused signature-based identification systems to fail at recognizing these new types of ransomware. Machine learning techniques have been used successfully to replace such signature-based ransomware detection systems, and certain models are capable of classifying ransomware into different family types [14].

Several works have been presented in recent years which document successful experiments using machine learning for ransomware classification. One recent work used a multi-layer perceptron (MLP) model (a type of deep learning) for classifying ransomware samples into different families [9]. They ran extensive experiments to select the best structure for the MLP model, and achieved a classification accuracy of 98%.

One study found that the network traffic that is created by ransomware activities can be used to identify the ransomware, using features of network packets such as query type and protocol type [15]. Two classifiers were set up for analyzing traffic, one at the flow level and the other at the individual packet level. They were able to achieve detection accuracy of 97.2% at the packet level and 97.08% at the flow level.

In [5], researchers attempted to detect the presence of ransomware in IoT devices by analyzing power consumption in the devices. Using a KNN machine learning model, they achieved an accuracy of 94.27% for classifying between ransomware and non-malicious processes.

Another study analyzed the API call sequences that are made when ransomware is executed, in order to differentiate between a ransomware executable and a benign executable [16]. Using a Simple Logistic machine learning algorithm, they were able to achieve an accuracy of 98.2%.

In another study for ransomware detection, network traffic was analyzed between a computer infected with ransomware and a Command and Control centre [17]. Using an RF classifier, they achieve an accuracy of just under 87%.

In 2019, Hanqi Zhang [18] performed text analysis to classify ransomware into different families. Patterns in the operation code of malware samples were extracted using text analysis and were used as features for machine learning models. Using these features, Hanqi Zhang [18] were able to achieve 91.43% accuracy in classifying the Wanacry family of ransomware using a RF machine learning model. In addition to this, 99.9% accuracy was recorded in discriminating between ransomware and trusted software.

Similarly, an experiment was conducted using text analysis in addition to reverse engineering [19]. In this experiment, they built datasets by performing reverse engineering on the binary code of ransomware samples and benign file samples. The average accuracy for all seven machine learning algorithms used in the study was greater than 90%. The highest accuracy was 97.95%, reported for the RF algorithm.

3 Methodology

This section will describe the process that was followed to build our five machine learning models, which are capable of classifying ransomware samples into one of six families. The six ransomware families in our dataset are: Cerber, CryptoWall, CTB-Locker, Locky, Sage, TeslaCrypt. We describe the process of identifying frequently occurring patterns in ransomware samples in order to have a feature set and feature values for each sample; we then explain how feature reduction was used to select the best features and reduce the overall dimensionality of our dataset.

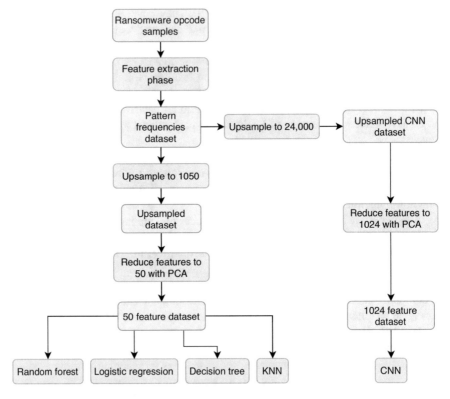

Fig. 1 Proposed methodology

Finally, we describe the design of our five machine learning models. An overview of the methodology can be seen in Fig. 1.

3.1 Feature Selection and Extraction

The provided ransomware samples consisted of text files for each sample, where each file is filled with operation code that would be executed by the ransomware. Each file falls into one of six possible ransomware family categories, as described above. For our experiments, each file is treated as a data point that needs to be classified. The features for each data point/file consist of the frequencies of different patterns; for example, if the first identified operation code pattern is "PUSH, PUSH, CALL", then the first feature value for each file will be the number of times which "PUSH, PUSH, CALL" appears in the file.

3.1.1 Pattern Selection

To obtain feature values for each ransomware sample, frequently occurring patterns needed to be extracted. A pattern consists of a subset of operation codes, and the frequency of patterns in each sample become the features values. Initially, all subsets of length at least two which occurred more than once in a given sample were considered as patterns. However, this resulted in thousands of unique patterns for only the first few files, and hundreds of files needed to be processed. To reduce the scale of our pattern selection problem, we assume that patterns which occur many times in a file are going to provide more insight for classifying a ransomware family as opposed to patterns which only occur twice. Following this logic, the minimum frequency to be considered a pattern was set to 50 (pattern must appear at least 50 times in a file to be recognized as a pattern), while the length of a pattern was limited to the range of 3–10 opcodes. This resulted in the total number of 6473 unique patterns, and therefore a new dataset was created where each sample/file has 6473 associated features. We will refer to this dataset as our "Pattern Frequencies" dataset.

3.1.2 Upsampling

In order to account for the fact that certain ransomware families had more samples than others, an upsampling technique was used on the "Pattern Frequencies" dataset so that the final dataset had an equal number of samples for each ransomware family. This would ensure that our models are not biased towards classifying samples as the majority class (the ransomware family for which we had the most samples). After upsampling, we had a dataset with 1050 samples, 175 for each family. We will refer to this dataset as our "Upsampled" dataset.

3.1.3 Upsampling for CNN

CNN models are designed to handle very large datasets, and as a result the "Upsampled" dataset does not contain enough data for optimal training of a CNN, despite having 1050 samples. Further upsampling was performed on the dataset until there was a total of 4000 samples per ransomware family, resulting in a final dataset with 24,000 samples. We will refer to this dataset as "Upsampled CNN".

3.1.4 Feature Reduction

Given that 6473 unique patterns were identified, feature reduction needed to be performed in order to reduce the overall dimensionality of our "Upsampled" and "Upsampled CNN" datasets. To achieve this, PCA was used. An implementation of PCA which is provided by the open source library scikit-learn was used [20]. PCA is a well-known machine learning technique which can reduce the number of features for a high-dimensional dataset [21]. PCA performs a transformation on the dataset where the features of the transformed dataset are ranked. The highest ranked feature corresponds to the feature which can account for the most variability in the dataset, the second ranked feature accounts for the second most variability, and so forth. Therefore, if n features are desired for analysis, the top n ranked features can be selected from PCA output. Through trial and error, it was determined that 50 was the optimal number of principal components to keep for our non-deep learning models. For our CNN model, 1024 features were kept from PCA because CNNs are well-suited to high-dimensional datasets (see Sect. 3.2.2 for more in-depth explanation of the input to our CNN model). We will refer to the 50-feature dataset as our "PCA" dataset, and the 1024-feature dataset as "CNN PCA".

3.2 Machine Learning Classifier Phase

Give different machine learning models were tested on the transformed output from PCA. KNN, RF, DT, and LG are all commonly used machine learning algorithms with fairly low complexities. The deep learning model, CNN, is more powerful but has higher computational complexity. All training and testing of models was performed on a 2.7 GHz Intel Core i5 processor.

3.2.1 KNN Model

Our KNN model was trained and tested on our "PCA" dataset, described in Sect. 3.1.4. An implementation of KNN which is provided by the open source library scikit-learn was used [20]. In KNN, a k value is selected, and each sample in the testing set is labelled based on the labels of the sample's k nearest neighbours in the training set [22]. That is, if k is equal to 6, and the 6 training samples around a particular testing sample are majority-labelled as class A, the testing sample will be labelled as A. The optimal k value for KNN on our dataset, discovered through trial and error, is 1. As a result, all testing with KNN was done with a k value of 1.

3.2.2 DT Model

Our DT model (obtained from [20]) was trained and tested on "PCA" dataset. The DT model designs a decision tree after learning about data, and then classifying a sample is as simple as tracing the decision tree based on sample characteristics. The maximum depth for the DT was set to 50 (training halts after tree reaches depth of 50).

3.2.3 RF Model

While the DT model creates a single decision tree, RF creates a "forest" of many decision trees. The RF model uses randomization to reduce the high variance and overfitting that is commonly exuded by a single decision tree. Algorithm for RF was taken from [20] and trained/tested on "PCA" dataset. The maximum number of trees in the RF model was set to 100.

3.2.4 LG Model

Logistic regression is a special type of regression analysis that is used for discrete classification purposes (regression is often used to estimate continuous values). Our LG model was taken from [20] and trained/tested on "PCA" dataset.

3.2.5 CNN Model

A CNN is a type of neural network, which is a form of deep learning. Deep learning is a subclass of machine learning which involves multiple neural networks working together for a single classification task [23]. The use of multiple neural networks causes deep learning algorithms to be extremely powerful, although they can be quite computationally expensive to run. The CNN used for this experiment is comprised of seven neural network layers, where the seventh layer is an output layer with six nodes (corresponding to our six classes of ransomware families). An implementation of CNN which is provided by the open source library keras was used [24]. Our CNN model was trained and tested on our "CNN PCA" dataset, described in Sect. 3.1.4.

CNNs were primarily invented for image classification, and so they expect each input sample to be in two dimensions. To accomplish this with our dataset, the best 1024 features were taken from PCA, so that each sample could be reshaped into a two-dimensional array of 32×32 (a common format size for images). This allows CNN to run on our dataset as though it were a dataset of image samples, where each image is 32 bytes by 32 bytes.

4 Experiments and Results

This section highlights the results of our experiments in which we attempt to classify ransomware samples into their respective family types, with the use of machine learning. We first describe the metrics used for evaluation of results, and then discuss the actual results from our experiments. We then compare our results to those in two similar papers, and finally discuss the ROC curves for our experiments.

4.1 Evaluation Measures

We will use metrics which are common for assessing our machine learning models, namely: TPR, FPR, accuracy, ROC curves, and AUROCs. Descriptions for TPR, FPR, and Accuracy can found in Eqs. (1), (2), and (3). The first three metrics will be discussed primarily in Sect. 4.2, while the ROC curve and AUROC will be discussed in Sect. 4.4.

4.1.1 TPR and FPR for Multi-Class Classification

Because TPR and FPR are typically calculated for binary classification (i.e. labels are either "Positive" or "Negative"), a slightly different approach needs to be taken for our multi-class classification problem. Our approach is to calculate TPR and FPR for each class individually, and then take the average of each. This is accomplished by outputting the results of our classification algorithms into confusion matrices (where the diagonal of a matrix contains the number of true positive samples for each class).

When viewing the results as a confusion matrix, it is possible to calculate the number of True Positives (TPs), True Negatives (TNs), False Positives (FPs), and False Negatives (FNs) for each class (see confusion matrix in Fig. 2). These values can then be used to calculate TPR and FPR for each class. For each class, TPR and FPR are calculated by denoting the class of interest as the "Positive" class, while the other classes are considered "Negative"; this is effectively what the "one-versus-all" approach is in machine learning when dealing with multiple class types. As such, TPR and FPR can be calculated for each ransomware family as if they had been classified in a one-versus-all fashion. For each ransomware family, a TP would be a sample of that ransomware family being correctly classified. The mean TPR value and mean FPR value can then be considered as the overall TPR and FPR.

$$TPR = TP/(TP + FN) \tag{1}$$

$$FPR = FP/(FP + TN) \tag{2}$$

$$Accuracy = (TP + TN)/(TP + FN + FP + TN) \tag{3}$$

Fig. 2 Confusion matrix

		Predicted	
		Positive	Negative
Actual	Positive	TP	FN
	Negative	FP	TN

4.2   Experiments and Results

We trained and tested our non-deep learning models on the "PCA" dataset, while our CNN model was trained and tested on the "CNN PCA" dataset; the "PCA" dataset contained 1050 samples and the "CNN PCA" dataset contained 24,000 samples, with an equal number of samples for each ransomware family. "PCA" contains the top 50 features outputted by PCA, and "CNN PCA" contains the top 1024 features outputted by PCA. A total of six different ransomware families are present in the dataset.

4.2.1   KNN Results

When running our KNN model, the ideal k value for our dataset (discovered after trial and error) was 1. This means that for every sample in the testing set, KNN classifies the sample according to the class label of whichever training sample is closest to the testing sample. Using tenfold cross validation, our model achieved an overall accuracy of approximately 92% when performing multi-class classification on ransomware samples. The overall TPR and FPR (based on the mean TPR and FPR for each class individually) was approximately 90% and 2%, respectively. Even when performing tenfold cross-validation, our KNN model requires less than a second to run.

4.2.2   DT Results

Our DT model received an accuracy of 89% using tenfold cross-validation, for multi-class classification. The average TPR for one-versus-all classification was approximately 90%, and the average FPR was approximately 2%. The DT model typically completed tenfold cross-validation just under 6 s.

4.2.3   RF Results

Our RF model received an accuracy of 92% using tenfold cross-validation, for multi-class classification. The average TPR for one-versus-all classification was approximately 91.5%, and average FPR was 1.5%. For this model, tenfold cross-validation is typically completed within 30 s (varies between 20 and 30 s).

4.2.4   LG Results

Our LG model was able to achieve an accuracy of 89% for multi-class classification. The average TPR reported for one-versus-all classification was 89%, and the average FPR was 1.5%. Our LG model typically completed tenfold cross-validation within 90 s.

4.2.5   CNN Results

Using tenfold cross-validation, CNN achieved an accuracy of 96%. The average values for TPR and FPR were 95% and 1%, respectively. When performing tenfold cross-validation, and training each CNN model for 5 epochs (5 rounds of training on each training set), our CNN model typically required at least 30 min to run. When setting the number of epochs to 20 for our CNN, using tenfold cross-validation we were able to achieve 96.9% accuracy; however, tenfold cross-validation with 20 epochs per fold took more than 3 h to run (on a 2.7 GHz Intel Core i5 processor).

4.2.6   Comparison of Our Models

Clearly, our CNN model outperforms all of our other models by up to 7%. This is likely due to the fact that CNN is a type of deep learning model and is therefore much more powerful than the non-deep learning algorithms. The increase in power is reflected in the increased runtime caused by CNN; on average, our CNN model typically takes at least 30 min to perform tenfold cross-validation on the ransomware dataset. As a result, it may be desirable to use a slightly less accurate machine learning algorithm that is capable of fast performance. Of the non-deep learning models, KNN is tied with RF for the second highest accuracy when classifying ransomware families. However, KNN typically takes less than a second to run whereas RF can take close to 30 s. As a result, if limited time is a factor, KNN is the best of our models for classifying ransomware families with fairly high accuracy. However, if time is not a factor, then CNN achieves the highest accuracy of our five models.

4.2.7   ROC Curves

An ROC curve is a common metric for evaluating the performance of machine learning classifiers, and is a great visual to see how well a particular model can separate two different classes (it is designed for binary classification) [12]. It is a plot of TPR against FPR at different threshold values, and the AUROC represents the overall performance of the model (AUROC = 1 is a perfect score).

As discussed in Sect. 4.1.1, metrics involving TPR and FPR needed to be calculated for each class individually, in a one-versus-all fashion. The ROC curves for classification of each ransomware family, for each of our five machine learning models, are shown in Figs. 3, 4, 5, 6, and 7. As shown in the legends, the AUROC for each family (where it says "area = ") is quite good for all five models (all are greater

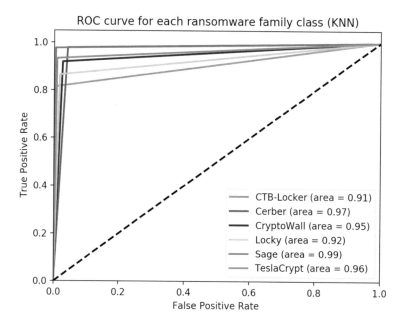

Fig. 3 ROC curves for KNN algorithm

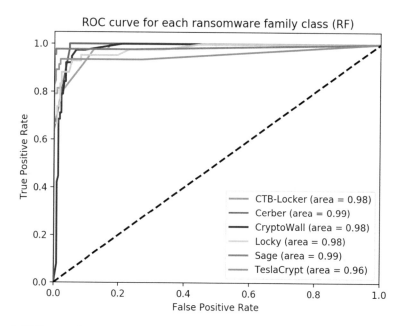

Fig. 4 ROC curves for RF algorithm

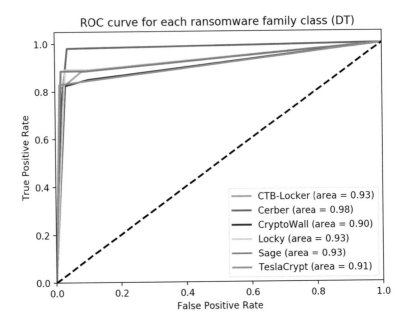

Fig. 5 ROC curves for DT algorithm

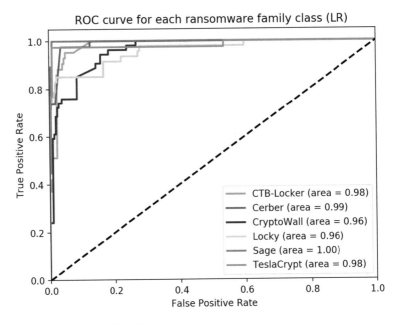

Fig. 6 ROC curves for LG algorithm

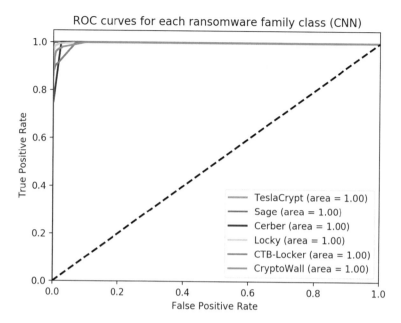

Fig. 7 ROC curves for CNN algorithm

than or equal to 0.9). It should be noted that all models have much higher accuracy when classifying in a one-versus-all fashion as opposed to multi-class classification.

The ROC curves from the CNN model (run with 5 epochs) are shown in Fig. 7. The AUROC values, displayed in the legend on the bottom-right of the image, are not actually 1 but are each extremely close to 1. Once again, this is a reflection of the fact that classifiers tend to perform better for binary classification (one-versus-all, in our case) as opposed to multi-class classification. This makes sense, as it is simply easier for a machine learning model to split a dataset into two different categories as opposed to 6.

Taking the average TPR and FPR for all five models resulted in the five ROC curves shown in Fig. 8. The average AUROC for CNN was not actually 1, but extremely close to 1. Again, the reason we do not see an AUROC closer to 0.96 (given that CNN accuracy was 96%) is because TPR and FPR values can only be calculated in a one-versus-all fashion.

4.3   Result Comparison

To validate our machine learning model results, we are highlighting two recent and relevant experiments performed on classification of ransomware families using machine learning.

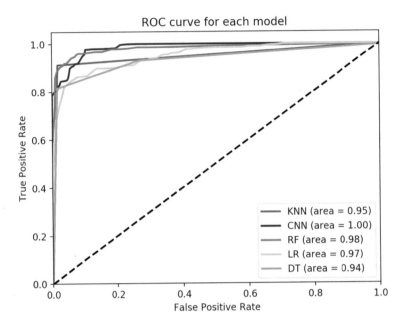

Fig. 8 Average ROC curve for all five models

In [18], an approach was proposed to classify ransomware families based on text analysis of operation codes, similar to the technique described in this paper. In [18], sequences in the operation codes were selected from malware samples using text analysis. The most prevalent opcode sequences were selected as features for the dataset. A metric called Term Frequency-Inverse Document Frequency (TF-IDF) was calculated for each feature, in order to select the best features for use with machine learning algorithms. The five algorithms which were used for the experiment are Decision Tree, Random forest, KNN, Naive Bayes and Gradient Boosting Decision Tree. The best accuracy reported in this work was 91.43%, using the Random Forest algorithm for classification of ransomware into family types.

In [11], machine learning models were built and tested on the same dataset of ransomware samples that was used in this paper. A method similar to ours was used to extract features from the ransomware samples: the authors used text analysis to obtain the most frequently occurring operation code patterns. Using these patterns as the features for each ransomware sample, several machine learning algorithms were designed to classify the samples into their respective family types. J48, Random Forest, Bagging and Multilayer Perceptron models were used. They were able to achieve an accuracy of 96.5% in classifying ransomware into ransomware families.

In this paper, we achieved 96% accuracy with our CNN model using 5 epochs per fold of cross-validation, and 96.9% accuracy using 20 epochs. While 96.9% is a slightly higher accuracy than what was reported in [11], more than 3 h was required to run our CNN on 20 epochs, 10 times in a row (for cross-validation).

5   Conclusion and Future Work

We were able to successfully design five different machine learning models for the purpose of classifying ransomware samples into their respective families. One model, KNN, achieved an overall accuracy of 92%, with a TPR of 90% and an FPR of 2%. KNN typically required less than a second to execute tenfold cross-validation. Our DT model achieved accuracy of 89%, with a TPR of 90% and an FPR of 2%. DT model typically required less than 6 s to run. RF model ran in less than 30 s and achieved an accuracy of 92%, with a TPR of 91.5% and FPR of 1.5%. LG model took close to 90 s to run, with an accuracy of 90%, a TPR of 89%, and FPR of 1.5%. The deep learning model, CNN, achieved an overall accuracy of 96% with a TPR of 95% and an FPR of 1%. 96.9% accuracy could be achieved with CNN if it was given more than 3 h to run (using 20 epochs). CNN typically required at least half an hour in order to execute tenfold cross-validation, using 5 epochs. The overall AUROC value for KNN was 0.95, 0.98 for RF, 0.97 for LR, and 0.94 for DT. The AUROC value for CNN was extremely close to 1. These results allowed us to perform a direct comparison of two different machine learning algorithms for classification of ransomware samples into their respective families. While CNN clearly offers higher accuracy than the other models, it is a highly complex algorithm which requires significant time and memory to execute. Of our five models, KNN is tied for the second-highest accuracy while having by far the lowest run time. The reasonably high accuracy that was achieved with KNN implies that it is a good algorithm to use if time is of the essence. If time is no factor and/or significant processing power is available, CNN offers very high accuracy; in fact, if time truly is not a factor then the number of epochs (iterations over training data) can be continually increased for CNN and the accuracy will most likely increase as well.

For future work, it would be ideal if the machine learning models designed in this paper could be tested on larger datasets of ransomware samples. While upsampling techniques provide a good method for increasing the size of a dataset to improve the performance of machine learning models, artificially creating data samples is not the same as collecting real-world samples of ransomware files. In particular, collection of more ransomware samples would be of great value for the CNN model because it performs so much better when it is provided with massive datasets. Furthermore, additional machine learning algorithms should be investigated which can achieve similar accuracy to CNN but with less computational complexity.

References

1. I. Nadir, T. Bakhshi, Contemporary cybercrime: A taxonomy of ransomware threats & mitigation techniques, in *2018 International Conference on Computing, Mathematics and Engineering Technologies (iCoMET)* (IEEE, Piscataway, 2018), pp. 1–7

2. O.M. Alhawi, J. Baldwin, A. Dehghantanha, Leveraging machine learning techniques for windows ransomware network traffic detection, in *Cyber Threat Intelligence* (Springer, Cham, 2018), pp. 93–106
3. R. Richardson, M.M. North, Ransomware: Evolution, mitigation and prevention. Int. Manag. Rev. **13**(1), 10 (2017)
4. A. Bhardwaj, V. Avasthi, H. Sastry, G. Subrahmanyam, Ransomware digital extortion: a rising new age threat. Indian J. Sci. Technol. **9**(14), 1–5 (2016)
5. A. Azmoodeh, A. Dehghantanha, M. Conti, K.-K.R. Choo, Detecting crypto-ransomware in IoT networks based on energy consumption footprint. J. Ambient Intell. Humaniz. Comput. **9**(4), 1141–1152 (2018)
6. N. Scaife, H. Carter, P. Traynor, K.R. Butler, Cryptolock (and drop it): stopping ransomware attacks on user data, in *2016 IEEE 36th International Conference on Distributed Computing Systems (ICDCS)* (IEEE, Piscataway, 2016), pp. 303–312
7. H. Daku, P. Zavarsky, Y. Malik, Behavioral-based classification and identification of ransomware variants using machine learning, in *2018 17th IEEE International Conference on Trust, Security and Privacy in Computing and Communications/12th IEEE International Conference on Big Data Science and Engineering (TrustCom/BigDataSE)* (IEEE, Piscataway, 2018), pp. 1560–1564
8. M. Nar, A.G. Kakisim, N. Çarkaci, M.N. Yavuz, I. Sogukpinar, Analysis and comparison of opcode-based malware detection approaches, in *2018 3rd International Conference on Computer Science and Engineering (UBMK)* (IEEE, Piscataway, 2018), pp. 498–503
9. R. Vinayakumar, K. Soman, K.S. Velan, S. Ganorkar, Evaluating shallow and deep networks for ransomware detection and classification, in *2017 International Conference on Advances in Computing, Communications and Informatics (ICACCI)* (IEEE, Piscataway, 2017), pp. 259–265
10. S. Homayoun, A. Dehghantanha, M. Ahmadzadeh, S. Hashemi, R. Khayami, K.-K.R. Choo, D.E. Newton, Drthis: Deep ransomware threat hunting and intelligence system at the fog layer. Futur. Gener. Comput. Syst. **90**, 94–104 (2019)
11. S. Homayoun, A. Dehghantanha, M. Ahmadzadeh, S. Hashemi, R. Khayami, Know abnormal, find evil: frequent pattern mining for ransomware threat hunting and intelligence. IEEE Trans. Emerg. Top. Comput. **8**, 341 (2017)
12. M. Hossin, M. Sulaiman, A review on evaluation metrics for data classification evaluations. Int. J. Data Min. Knowl. Manag. Process **5**(2), 1 (2015)
13. C. Srinivasan, Hobby hackers to billion-dollar industry: the evolution of ransomware. Comput. Fraud Secur. **2017**(11), 7–9 (2017)
14. S. Maniath, A. Ashok, P. Poornachandran, V. Sujadevi, A.P. Sankar, S. Jan, Deep learning LSTM based ransomware detection, in *2017 Recent Developments in Control, Automation & Power Engineering (RDCAPE)* (IEEE, Piscataway, 2017), pp. 442–446
15. A.O. Almashhadani, M. Kaiiali, S. Sezer, P. O'Kane, A multi-classifier network-based crypto ransomware detection system: A case study of locky ransomware. IEEE Access **7**, 47053–47067 (2019)
16. Z.-G. Chen, H.-S. Kang, S.-N. Yin, S.-R. Kim, Automatic ransomware detection and analysis based on dynamic API calls flow graph, in *Proceedings of the International Conference on Research in Adaptive and Convergent Systems* (2017), pp. 196–201
17. G. Cusack, O. Michel, E. Keller, Machine learning-based detection of ransomware using SDN, in *Proceedings of the 2018 ACM International Workshop on Security in Software Defined Networks & Network Function Virtualization* (2018), pp. 1–6
18. H. Zhang, X. Xiao, F. Mercaldo, S. Ni, F. Martinelli, A.K. Sangaiah, Classification of ransomware families with machine learning based on n-gram of opcodes. Futur. Gener. Comput. Syst. **90**, 211–221 (2019)
19. S. Poudyal, K.P. Subedi, D. Dasgupta, A framework for analyzing ransomware using machine learning, in *2018 IEEE Symposium Series on Computational Intelligence (SSCI)* (IEEE, Piscataway, 2018), pp. 1692–1699

20. F. Pedregosa, G. Varoquaux, A. Gramfort, V. Michel, B. Thirion, O. Grisel, M. Blondel, P. Prettenhofer, R. Weiss, V. Dubourg, J. Vanderplas, A. Passos, D. Cournapeau, M. Brucher, M. Perrot, E. Duchesnay, Scikit-learn: Machine learning in Python. J. Mach. Learn. Res. **12**, 2825–2830 (2011)
21. F.P. Shah, V. Patel, A review on feature selection and feature extraction for text classification, in *2016 International Conference on Wireless Communications, Signal Processing and Networking (WiSPNET)* (IEEE, Piscataway, 2016), pp. 2264–2268
22. Z. Deng, X. Zhu, D. Cheng, M. Zong, S. Zhang, Efficient KNN classification algorithm for big data. Neurocomputing **195**, 143–148 (2016)
23. J. Schmidhuber, Deep learning in neural networks: An overview. Neural Netw. **61**, 85–117 (2015)
24. F. Chollet et al., Keras (2015). https://keras.io/getting_started/faq/#how-should-i-cite-keras

Scalable Fair Clustering Algorithm for Internet of Things Malware Classification

Zibekieni Obuzor and Adesola Anidu

1 Introduction

Machine learning algorithms have been applied to solve various real-world problems such as malware classification and detection [1–4], malware hunting [5–7], image processing [8], predictions, and etc. [9]. However, there are concerns that results obtained from machine learning algorithms are not always fair [10]. This brings the notion to investigate what it means for an algorithm to be fair and how this fairness can be achieved when designing algorithms [10]. The interpretation of fairness in the machine learning community is the use of disparate impact notion of fairness [11]. This has been used by [12] in designing fair classification. Also, [13] has also employed the same method for multi-winner voting with fairness constraints. The first application of the disparate notion of fairness to the clustering problem was done by [10, 14].

Clustering is a technique in machine learning that involves the grouping of data points and classifying the data points into a specific group [15, 16]. Data points in each group are similar to each other and differ from the data points in other clusters. There are different clustering methods, and each may generate different clusters using the same set of data points. The clustering algorithm is an unsupervised learning algorithm with many real-world applications such as being used as a preprocessing step in solving classification problems [17]. Fairness in clustering is an instance where each cluster has approximately equal representation of each protected class though there are instances where a point is not assigned to the nearest cluster center [14]. Minimal sets that satisfy this fair representation are known as fairlets. Fairlets were introduced to preserve the clustering objectives [14]. As a

Z. Obuzor (✉) · A. Anidu
School of Computer Science, University of Guelph, Guelph, ON, Canada
e-mail: zobuzor@uoguelph.ca; aanidu@uoguelph.ca

© Springer Nature Switzerland AG 2021
K.-K. R. Choo, A. Dehghantanha (eds.), *Handbook of Big Data Analytics and Forensics*, https://doi.org/10.1007/978-3-030-74753-4_18

result, any fair clustering problem is decomposed to finding good fairlets initially, then the existing traditional clustering algorithm is used [14]. The different distance metrics used with the clustering method include k-means, k-median and k-center objectives.

Big data and scalability are serious concern that should be addressed while machine learning models are proposed for real-world applications [15, 18–22]. Scalable fair clustering is a fair variant of the k-median problem introduced by [14]. The main objective of a k-median problem is to find k centers C with each input point assigned to one of the centers in C in a way that the average distance of data points to the center of the cluster is minimized [23]. In the fair variant of the k-median, the minimization of the average distance objective is the main aim while making sure that all clusters have an "approximately equal" number of points of each class [23]. This algorithm runs in nearly linear time which indicates an improvement over the algorithm presented by [14] which runs in super-quadratic time. While this is applaudable, it is important to investigate the impact of this fairness on the accuracy of clustering a given dataset.

Previous research studies on fair clustering algorithms have emphasized on ensuring fair clusters and improving running time [24–34]. None of these studies has focused on the outcome of the fair clustering algorithm on accuracy. At the time of this study, this is the first evaluation to be implemented on the accuracy of a fair clustering algorithm. This fair clustering algorithm by [11] will be employed in clustering and detecting malware in the Internet of Things (IoT) devices. IoT devices have become popular and are prone to malware attacks because of the rate at which these devices are developed [35, 36]. Malware detection can either be done by dynamic or static analysis [37]. Dynamic analysis involves the execution of a program in an emulator or instrumented hardware to extract characteristics actions executed by the program [38] while static analysis involves the disassembling of program binary to extract the features such as strings, opcodes and so on [39].

IoT malware dataset which is made up of goodware (benign) and malware opcodes extracted by static malware analysis will be used for this experiment. Opcodes are mnemonic for operational codes. Studies have shown that the opcodes feature extraction method is more efficient in detecting malware. According to [40] opcodes reveal a lot of statistical differences between legitimate software and malware. A recent study by [41] in the use of opcodes for detection of polymorphic IoT malware using Support Vector Machine (SVM), K-Nearest Neighbor (KNN), Multilayer Perceptron (MLP), Adaboost, Decision Tree and Random Forest yielded an accuracy of 99%, thus, buttressing the findings of 37. The accuracy rate of the implementation of a fair clustering algorithm on malware opcodes for clustering and detection of IoT malware will be compared with the most recent study on the detection of IoT malware using opcodes presented by [41].

The next section briefly describes the research work related to the topic. Section 3 describes the methodology of the study. In Sect. 4, the experiment and results are presented. Lastly, Sect. 5 concludes the paper and presents the future research direction.

2   Literature Review

In this section, a review of previous research work related to the application of machine learning algorithms for securing IoT, the use of opcodes for IoT malware detection and, the use of clustering algorithms in malware classification/detection is carried out.

Various machine learning based solutions [42] have been employed for securing IoT devices. According to [42], machine learning methods have been used to detect new attacks and can provide potential security protocols that can be utilized in IoT devices, thus, making them more accessible and reliable. A deep and scalable unsupervised machine learning system was presented by [17] for detecting cyber-attack in large-scale smart grids. The system utilizes feature extraction using symbolic dynamic filtering (SDF) for the reduction of computational load and casual interactions between subsystems were also discovered. An accuracy of 99% was obtained as results from simulations on IEEE39, 118, and 2848 bus systems. Jahromi et al. [43] proposed a modified Two-hidden-layered Extreme Learning Machine (TELM) for the detection of malware. The proposed method employs the dependency of malware sequence elements through the avoidance of backpropagation when neural networks are trained. The method accelerates the training and detection steps for malware hunting when compared with the Long Short-Term Memory (LSTM) and Convolutional Neural Network (CNN). An accuracy rate of 99.65% was achieved when tested using an IoT-specific dataset and the method can be used on all platforms for malware analysis. Also, [26] presented a deep learning method for detecting crypto-mining malware utilizing static and dynamic analysis. 1500 Portable Execution (PE) which is made up of system call events were captured for dynamic analysis. When LSTM, CNN, and Attention-based LSTM (ATT-LSTM) approaches were implemented on sequences of system call events, an accuracy of 95% and 99% were obtained for static and dynamic analysis.

Also, [44] presented a novel light-weight method for the detection of DDoS malware in the IoT environment is presented. The extraction of one-channel gray-scale images converted from binaries is done. A light-weight convolutional neural network is utilized for classifying the IoT malware families. An accuracy of 94.0% was achieved in classifying goodware and DDoS malware and an accuracy of 81.8% was obtained for the classification of goodware and two main malware families. A malware detection method using family behavior graph is presented by [45]. Malware behaviors are represented as dependency graphs. Common behavior graphs are extracted to represent the behavioral features of the malware family. Malicious codes are detected using a graph matching algorithm that is dependent on the maximum weight subgraph. Results from experiments showed that the method has a high detection rate, low false-positive rate and can detect malware variants. A graph-based model called the System-call Dependency Graphs (ScD-graphs) that utilizes relations between system calls for the detection and classification of malware is presented by [46]. Results from evaluation show that it can be used for malware detection and classification. Also, a novel byte-level method for the detection of malware using audio signal processing techniques is proposed by [47] Conversion of the program's bytes into meaningful audio signals is carried out, then the construction of a machine learning music classification model from audio signals is done using Music Information Retrieval (MIR) techniques to detect new and unseen instances.

Furthermore, a model that detects malware using API call sequences, text mining, and topic modeling is proposed by [48]. Results from experiments conducted using Decision Tree and Support Vector Machine yielded good results. Behavioral differences are used to distinguish between malware and benign programs using data mining techniques. Results from the experiment showed that the detection rate is 95% with 80 attributes [49]. Profile Hidden Markov Models (PHMMs) were trained using API call sequences. Results showed that dynamic analysis produces better results [50]. Two machine learning methods for static analysis of mobile applications is presented by [2]. A source code-based analysis using a bag of words representation model and permission-based approaches are presented. F-measure of 95.1% and 89% were achieved for the source code-based method and permissions-based method, respectively. Damshenas et al. [1] presented M0Droid, a novel Android behavioral-based malware detection approach. A signature is generated for each application based on the system request of the application. The signature generated is normalized to improve accuracy. Malware identification is done by checking the degree of similarity of behavior to the existing blacklist of malware signatures using Spearman's rank correlation coefficient. The detection rate of 60.16% with 0.4% false-negatives and 39.43% false positives were achieved at a threshold value of 0.90 when experiments were run using M0Droid on Genome dataset and APK submissions of Android client agent. Huda et al. [51] proposed a hybrid framework for malware detection using hybrids of Support Vector Machine wrapper and Maximum-Relevance-Minimum-Redundancy Filter. Application Program Interface (API) call statistics are used as features of the malware. The performance of this proposed method is compared using binary logistic regression. Results from experiments performed showed that the proposed hybrid framework performs better in identifying malware.

Operational Codes (Opcodes) from programs have been introduced as a reliable feature for identifying and detecting malware using machine learning in devices. Header information, ByteCodes, attacker's intent, API calls, and permissions have been combined with opcodes to create a multi-view learning method for the detection of malicious programs [41]. Results from experiments conducted across various platforms such as IoT, Android, and Windows showed a high accuracy rate with a low rate of false-positive [52]. Malware detection using a control flow-based opcode behavior was presented by [53]. Executable opcode behaviors were extracted using a control flow-based method. A control flow graph for the program is created to determine the opcode behavior of the program from the execution paths when the graph is traversed. Results presented showed a higher accuracy rate and a lower false-positive rate when compared with other text-based detection methods. Also [39] employed a fuzzy and fast fuzzy pattern tree method on the program's opcodes for categorizing and detecting malware with very high accuracy. Hashemi et al. [54] presented a new malware detection method based on opcodes in an executable file. A graph of opcode within an executable file is generated. The graph is thereafter embedded into the eigenspace using the Power Iteration method. Executable files are presented as a linear combination of eigenvectors proportionate to their respective eigenvalues. When evaluated using SVM and KNN, the proposed method has a high detection rate, low false-positive rate, and acceptable computational complexity.

Furthermore, [55] proposed an IoT malware threat hunting using a deep recurrent neural network-based approach. Opcodes were extracted to build the datasets for malware and benign ware samples. For each opcode sample, feature vector files were created. The LSTM was used to design the deep learning structure for the detection of the IoT malware samples based on the opcodes' sequence. Google Tensor Flow was used as a backend structure and Scikit-learn as the machine learning library for evaluation. Detection accuracy of 98% was achieved when evaluated with ARM-based IoT applications' execution codes. Azmoodeh et al. [56] used Convolutional Network for detection of malware in IoT and Internet of Battlefield Things (IoBT) using opcodes. Selected opcode sequence was used as a feature for the classification task after which a graph of features for each sample was created. Malware classification was carried out using deep Eigenspace learning. The approach achieved an accuracy of 98.37% in malware detection and a precision rate of 98.59%. Also, [41] used a sequential pattern mining technique for the detection of the most frequent opcode sequences in malicious IoT applications. Sequential pattern mining algorithms were used in the extraction of sub-sequences embedded in the text given which is based upon a support value and a user-specified threshold. Sequential pattern mining was combined with other machine learning techniques for the classification of IoT goodware, malware, and polymorphic malware samples. The dataset used was made up of 269 IoT good ware and 247 malware samples. 36 features were detected because of frequency in patterns of malware opcodes and the division to transitional and atomic types. An accuracy and f-measure greater than 99% were achieved in detecting IoT malware from benign samples using KNN, MLP, SVM, random forest, Adaboost, and Decision Tree machine learning classifiers.

Malware detection can also be achieved using clustering algorithms. Clustering is an unsupervised machine learning algorithm that partitions a given dataset into clusters/groups based on pre-defined distance measured in a manner that data points closed to each other are in a cluster and data points in different clusters are far from each other [57]. The use of any clustering algorithm either traditional or modern in malware detection is based on the features extracted and the underlying feature distribution [58]. Visaggio et al. [59] classified malware based on k-means and Expectation Maximization (EM) clustering algorithms. The experiment was conducted on 8000 malware samples with the number of clusters varied from 2 to 10 and the number of dimensions varied from 2 to 5. The EM performed better than k-means. Also, the result showed that clustering can be used in detecting new malware samples. Stamp et al. [60] also presented a paper where the Hidden Markov

Model (HMM) analysis and clustering techniques were used to classify malware. The malware families were trained using HMM and the resulting models were used to score samples for the malware families. K-means and EM clustering algorithms were applied to cluster the malware samples based on the HMM scores. The results obtained are comparable with when the Support Vector Machine (SVM) approach is used. Fuzzy hashing algorithms have been utilized for malware clustering analysis and can be used in malware similarity analysis [61]. Alaeiyan et al. [16] presented a multi-label fuzzy clustering system for malware attack attribution. In the proposed method, opcode frequencies were employed as feature spaces for the classification of different malware families. Samples from VirusShare, BIG2015, and RansomwareTracker were used to test the proposed method and an accuracy of 94.66%, 97.56%, and 94.26% were obtained, respectively.

Despite the promising results obtained from using clustering algorithms, there are concerns that existing clustering algorithms do not exhibit fairness. Several research works are ongoing in the field. Some of the proposed variants for fair clustering are reviewed below.

A fair clustering problem under the k-median and k-center objectives are formulated by [14]. In their work, they demonstrated that any fair clustering problem can be decomposed into finding good fairlets first and then processing using traditional clustering algorithms. The results from the experiment indicate that traditional algorithms produce unfair clusters and that the algorithm developed guarantees fair clusters. Another variant is the fair variant of the near neighbor problem presented by [62]. The aim is to pre-process the points in a manner that given any query point q, any point within the r-neighborhood of the query has the same probability of being picked as the near neighbor. This indicates that the Locality-Sensitive Hashing (LSH) based algorithm can also be made fair without a loss in efficiency. The results showed that the algorithm works well in producing an empirical distribution that is closer to the uniform distribution. Fair clustering for multiple sensitive attributes is presented by [63]. An outline of a computational notion of fairness is implemented with a cluster coherence objective to develop the FairKM clustering method. The results from the experiment showed that the clusters generated using the FairKM algorithm are significantly better for fair representation of sensitive groups and the quality of clustering when compared to the clusters from other fair clustering methods. Also, [64] presented an algorithm that minimizes the classical clustering cost with the condition that there is no over-representation of color in each cluster. Results from experiments conducted on real-world data showed that the algorithm can find good clusters without over-representation. Bercea et al. [65] presented a way by which fair clustering can be established when the centers are already fixed. A 5-approximation for the fair k-center problem with protected classes is given and r relaxed fairness notion is proposed where bicriteria constant-factor approximations for all classical clustering objectives k-supplier, k-center, k-median, k-means, and facility location are given. Furthermore, a new meta-algorithm for classification is presented by [66]. A large class of fairness constraint is taken as input considering non-disjoint sensitive attributes. Empirical results obtained showed that the algorithm can achieve near-perfect fairness with respect to various fairness metrics. Chen et al. [67] considered the problem of proportional centroid clustering. Fairness for clustering n points with k centers is defined as proportionality that means that any n/k points can form their cluster provided there exists another center which is closer in distance for all n/k points. Algorithms are analyzed for computation, optimization, and auditing of proportional solutions.

3   Methodology

This section presents an overview of the steps taken in preparing the dataset, implementing the algorithm, and evaluating the result.

3.1   Step 1: Transformation of the Dataset

As with any machine learning problem, there is a need to prepare the dataset. This is done to transform the data into the right format. The raw dataset consists of the program opcode for 268 different samples of goodware (benign) and 244 different samples of malware stored as text files. The total number of text files in the raw dataset is 512. These 512 text files must be stored as a sequence of comma-separated values (csv).

Extraction of the vocabulary of all opcodes was done using a program written in Python programming language. The vocabulary was used to create the dictionary of words with corresponding frequency in each opcode in each of the text files. The dictionary of words which is made up of the opcodes and their frequency transformed to csv by exporting it to Microsoft Excel. There are 681 possible features consisting of 305 unique opcodes and 376 names of the different application processors embedded in each text file.

3.2   Step 2: Pre-processing of the Dataset

We developed a data pre-processing module to pre-process the data based on the requirements stated in the algorithm. The feature selection is done using the Principal Component Analysis (PCA) algorithm. The pre-processing module does the following:

- Loads the dataset including the target column (target column contains either "malware" or "benign")
- Performs a feature selection using PCA for the different number of features to be considered

- Replaces strings (i.e. malware and benign) within the target column with 0's and 1's
- Moves target column to the first column to comply with scalable fair clustering requirement

3.3 Step 3: Run the Clustering Algorithm

The steps taken to derive k-clusters using fair clustering algorithm [23] are summarized below:

- Computation of an approximately optimal (r, b) – fairlet decomposition which is represented by the input point set P
- Clustering of the (r, b) -fairlets produced in the previous step into k clusters using the k-median algorithm
- The resulting clustering is then extended to the whole dataset by assigning each data point to the cluster that contains its fairlet center thus resulting in the final fair clustering

3.4 Step 4: Interpret the Results

To evaluate the performance of the fair clustering algorithm in classifying the IoT malware opcodes dataset, the following criteria were used:

Accuracy can be defined as the number of samples which is detected correctly by the classifier divided by the total number of samples

$$Accuracy\ (\%) = \frac{TP + TN}{TP + TN + FP + FN} * 100$$

where

- True Positive (TP): indicates the number of true positives
- True Negative (TN): indicates the number of true negatives
- False Positive (FP): indicates the number of false positives
- False Negative (FN): indicates the number of false negatives

Precision is the ratio of predicted malware samples correctly labelled as a malware.

$$Precision = \frac{TP}{TP + FP}$$

Recall is the ratio of malware samples correctly predicted

$$Recall = \frac{TP}{TP + FN}$$

F-Measure is the harmonic mean of recall and precision

$$F - Measure = 2 * \frac{precision * recall}{precision + recall}$$

Fairlet decomposition cost is the cost of a fairlet (cluster) using the fairlet decomposition algorithm. Cost here denotes the total distances of the points to their cluster/fairlet centroids.

Fairlet decomposition runtime is the time taken for the fairlet decomposition algorithm to run in seconds.

4 Experiment

Here, the performance of the fair clustering algorithm is evaluated using the IoT malware opcodes dataset developed by researchers at the Cybersecurity laboratory of the University of Guelph. The dataset has 512 samples consisting of 268 goodware (benign) and 244 malware samples. Each sample consists of a list of opcodes extracted using static malware analysis. The experiment was done using Python 3.6, the code invokes MATLAB for the k-median clustering algorithm. For the algorithm to work efficiently, the following parameters were computed:

- **Balance:** p/q which can be less than or equal to 1 (for this experiment, the balance used is 0.8)
- **p, q:** two integers that define the desired cluster balance (for this experiment, $p = 4$, $q = 5$)
- **k:** number of output clusters (k in k-median) (k is varied)

Initially, we tested the fair clustering algorithm to know if the running time will be near-linear as reported. This was done by reproducing the experiment with one of the datasets used by [23] and comparing the results. The Diabetes dataset from the UCI Machine Learning Repository was used which represents information and outcome of patients with Diabetes in clinical care for 10 years from 130 US hospitals.

Figure 1 shows the result obtained from the original paper while Fig. 2 shows the result we obtained after using the same algorithm and dataset. When compared, it shows the result we got is comparable (nearly linear) to the result reported. We believe the little discrepancy in runtime is due to the processor speed of the different machines used. With the certainty of algorithm producing fair clusters, we proceeded to use the algorithm with the IoT malware opcodes dataset.

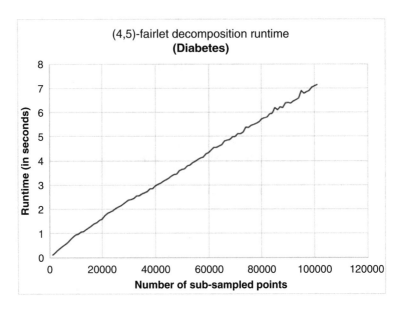

Fig. 1 Base paper result using Diabetes dataset [11]

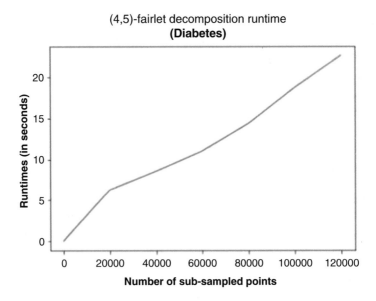

Fig. 2 Our result using Diabetes dataset

Table 1 TP, TN, FP, FN, Precision, Recall and F-Measure for each cluster

		5 features	10 features	20 features	30 features
2 clusters	TP	184	183	183	183
	TN	202	204	205	205
	FP	84	85	85	85
	FN	42	40	39	39
	Precision	0.6866	0.6828	0.6828	0.6828
	Recall	0.8142	0.8206	0.8243	0.8243
	F-Measure	0.7449	0.7454	0.7469	0.7469
5 clusters	TP	69	69	**220**	**220**
	TN	145	146	**209**	**209**
	FP	199	199	**48**	**48**
	FN	99	98	**35**	**35**
	Precision	0.2575	0.2575	0.8209	0.8209
	Recall	0.4107	0.4132	0.8627	0.8627
	F-Measure	0.3165	0.3172	0.8413	0.8413
10 clusters	TP	85	68	191	189
	TN	118	192	132	132
	FP	183	200	77	79
	FN	126	52	112	112
	Precision	0.3172	0.2537	0.7127	0.7052
	Recall	0.4028	0.5667	0.6304	0.6279
	F-Measure	0.3549	0.3505	0.6690	0.6643
15 clusters	TP	139	229	156	232
	TN	108	116	95	145
	FP	129	39	112	36
	FN	136	128	149	99
	Precision	0.5187	0.8545	0.5821	0.8657
	Recall	0.5055	0.6415	0.5115	0.7009
	F-Measure	0.5120	0.7328	0.5445	0.7746
20 clusters	TP	229	221	120	217
	TN	124	135	128	101
	FP	39	47	148	51
	FN	120	109	86	143
	Precision	0.8545	0.8246	0.4478	0.8097
	Recall	0.6562	0.6697	0.5825	0.6028
	F-Measure	0.7423	0.7391	0.5063	0.6911

5   Results

Using Table 1, the accuracy rate is computed and plotted as seen in Fig. 3. The highlighted area in Table 1 represents the TP, TN, FN, and FP that gave the highest accuracy rate when computed. The highest accuracy of 83.79% was obtained when 5 clusters were created using 20 and 30 features respectively as seen in Fig. 3 above.

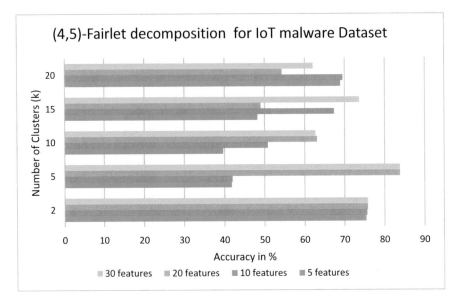

Fig. 3 Accuracy (%) for classifying IoT malware dataset

Also, from Fig. 4 above, the cost of fairlet decomposition increased as the number of features increased for 2, 5, and 10 clusters. However, this proportionality did not continue as the cluster size is increased for 15 and 20 clusters. Furthermore, from Fig. 5, the fairlet decomposition time for the dataset is nearly linear as the number of features is increased for each cluster.

The balance for the clusters can be varied to obtain fair clusters. We used other different balance values 0.5 and 0.25 respectively, to investigate whether there will be an improvement in the accuracy. However, we did not notice any improvement in accuracy. Based on the outcome, we conclude that the balance does not affect the accuracy but the fairness of clusters.

The accuracy result obtained is low when compared to the most recent studies about using IoT malware opcodes for classifying malware. Darabian et al. [41] obtained an average accuracy of 99% with the different machine learning algorithms implemented while we obtained 83.79%. This low accuracy result could be because accuracy was not put into consideration in the algorithm rather, the fairness of clusters, and reduction in runtime. Also, according to [68] there is a tradeoff between fairness and accuracy. The more the fairness, the more the random-level of accuracy which could also account for the low accuracy achieved. However, based on the results obtained in Fig. 5, the fairlet decomposition time is nearly linear which satisfies the objective of the fair clustering algorithm.

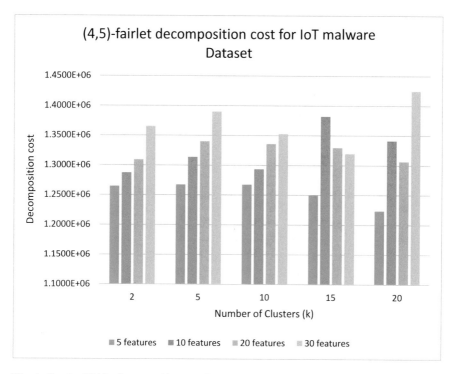

Fig. 4 Graph of fairlet decomposition cost for IoT malware dataset

6   Conclusion and Future Work

The fair clustering algorithm accuracy rate for the classification of IoT malware opcodes is low when compared to other machine learning algorithms. An accuracy that is higher than 83.79% has been obtained using other machine learning algorithms such as SVM. However, the main purpose of this fair clustering algorithm is to reduce the fairlet decomposition time from super-quadratic time to nearly linear time which was achieved. The main objective of the fair clustering algorithm we reproduced was fairness in its clustering and accomplishment of this fairness in nearly linear time. We were able to achieve this nearly linear time with the IoT malware opcodes dataset.

In future, performance evaluation of this dataset with other variants of the fair clustering algorithm can be done to further investigate the impact of the fairness of algorithm on accuracy. Furthermore, this fair clustering algorithm can be combined with other machine learning algorithms to test whether there will be an improvement in the performance of such algorithms.

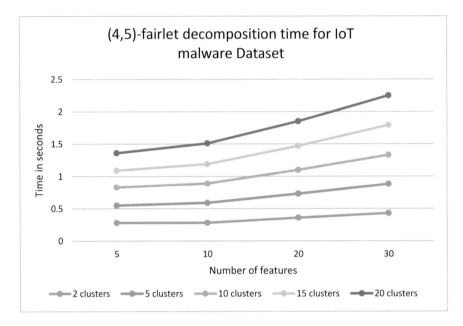

Fig. 5 Graph of the decomposition time of IoT malware dataset

References

1. M. Damshenas, A. Dehghantanha, K.-K.R. Choo, R. Mahmud, M0droid: An android behavioral-based malware detection model. J. Inf. Priv. Secur. **11**(3), 141–157 (2015)
2. N. Milosevic, A. Dehghantanha, K.-K.R. Choo, Machine learning aided android malware classification. Comput. Electr. Eng. **61**, 266–274 (2017)
3. A. Azmoodeh, A. Dehghantanha, M. Conti, K.-K.R. Choo, Detecting crypto-ransomware in IoT networks based on energy consumption footprint. J. Ambient. Intell. Humaniz. Comput. **9**(4), 1141–1152 (2018)
4. M. Saharkhizan, A. Azmoodeh, A. Dehghantanha, K.-K.R. Choo, R.M. Parizi, An ensemble of deep recurrent neural networks for detecting IoT cyber attacks using network traffic. IEEE Internet Things J. **7**(9), 8852–8859 (2020). https://doi.org/10.1109/jiot.2020.2996425
5. H.H. Pajouh, R. Javidan, R. Khayami, D. Ali, K.-K.R. Choo, A two-layer dimension reduction and two-tier classification model for anomaly-based intrusion detection in IoT backbone networks. IEEE Trans. Emerg. Top. Comput. **7**, 314–323 (2016)
6. S. Homayoun, A. Dehghantanha, M. Ahmadzadeh, S. Hashemi, R. Khayami, Know abnormal, find evil: Frequent pattern mining for ransomware threat hunting and intelligence. IEEE Trans. Emerg. Top. Comput. **6750**, 1–11 (2017)
7. M. Saharkhizan, A. Azmoodeh, H. HaddadPajouh, A. Dehghantanha, R.M. Parizi, G. Srivastava, A hybrid deep generative local metric learning method for intrusion detection, in *Handbook of Big Data Privacy*, (Springer, Cham, 2020), pp. 343–357. https://doi.org/10.1007/978-3-030-38557-6_16
8. K. Bolouri, A. Azmoodeh, A. Dehghantanha, M. Firouzmand, Internet of things camera identification algorithm based on sensor pattern noise using color filter array and wavelet transform, in *Handbook of Big Data and IoT Security*, (Springer, Cham, 2019), pp. 211–223. https://doi.org/10.1007/978-3-030-10543-3_9

9. A. Yazdinejad, R.M. Parizi, A. Dehghantanha, K.-K.R. Choo, Blockchain-enabled authentication handover with efficient privacy protection in SDN-based 5G networks. IEEE Trans. Netw. Sci. Eng. (2019). https://doi.org/10.1109/TNSE.2019.2937481

10. S. Bera, D. Chakrabarty, N. Flores, M. Negahbani, Fair algorithms for clustering, in *Advances in Neural Information Processing Systems*, (The MIT Press, Cambridge, MA, 2019), pp. 4954–4965

11. H. HaddadPajouh, R. Khayami, A. Dehghantanha, K.-K.R. Choo, R.M. Parizi, AI4SAFE-IoT: An AI-powered secure architecture for edge layer of Internet of things. Neural Comput. Applic. **32**(20), 16119–16133 (2020). https://doi.org/10.1007/s00521-020-04772-3

12. M. Feldman, S.A. Friedler, J. Moeller, C. Scheidegger, S. Venkatasubramanian, Certifying and removing disparate impact, in *Proceedings of the 21th ACM SIGKDD International Conference on Knowledge Discovery and Data Mining*, (2015), pp. 259–268

13. L.E. Celis, L. Huang, N.K. Vishnoi, Multiwinner voting with fairness constraints. arXiv Prepr. arXiv1710.10057 (2017)

14. F. Chierichetti, R. Kumar, S. Lattanzi, S. Vassilvitskii, Fair clustering through fairlets, in *Advances in Neural Information Processing Systems*, (The MIT Press, Cambridge, MA, 2017), pp. 5029–5037

15. A. Azmoodeh, A. Dehghantanha, R.M. Parizi, S. Hashemi, B. Gharabaghi, G. Srivastava, Active spectral botnet detection based on eigenvalue weighting, in *Handbook of Big Data Privacy*, (Springer, Cham, 2020), pp. 385–397. https://doi.org/10.1007/978-3-030-38557-6_19

16. M. Alaeiyan, A. Dehghantanha, T. Dargahi, M. Conti, S. Parsa, A multilabel fuzzy relevance clustering system for malware attack attribution in the edge layer of cyber-physical networks. ACM Trans. Cyber-Physical Syst. **4**(3), 1–22 (2020)

17. H. Karimipour, A. Dehghantanha, R.M. Parizi, K.-K.R. Choo, H. Leung, A deep and scalable unsupervised machine learning system for cyber-attack detection in large-scale smart grids. IEEE Access **7**, 80778–80788 (2019)

18. A. Yazdinejad, R.M. Parizi, A. Dehghantanha, K.-K.R. Choo, P4-to-blockchain: A secure blockchain-enabled packet parser for software defined networking. Comput. Secur. **88** (2020). https://doi.org/10.1016/j.cose.2019.101629

19. A. Al-Abassi, H. Karimipour, A. Dehghantanha, R.M. Parizi, An ensemble deep learning-based cyber-attack detection in industrial control system. IEEE Access **8**, 83965–83973 (2020)

20. M. Amrollahi, A. Dehghantanha, R.M. Parizi, A survey on application of big data in fin tech banking security and privacy, in *Handbook of Big Data Privacy*, (Springer, Cham, 2020), pp. 319–342

21. A. Azmoodeh, A. Dehghantanha, Big data and privacy: Challenges and opportunities, in *Handbook of Big Data Privacy*, (Springer, Cham, 2020), pp. 1–5. https://doi.org/10.1007/978-3-030-38557-6_1

22. J.C. Cabello, H. Karimipour, A.N. Jahromi, A. Dehghantanha, R.M. Parizi, Big-data and cyber- physical systems in healthcare: Challenges and opportunities, in *Handbook of Big Data Privacy*, ed. by K.-K. R. Choo, A. Dehghantanha, (Springer, Cham, 2020)

23. A. Backurs, P. Indyk, K. Onak, B. Schieber, A. Vakilian, T. Wagner, Scalable fair clustering. arXiv Prepr. arXiv1902.03519 (2019)

24. A. Yazdinejad, H. HaddadPajouh, A. Dehghantanha, R.M. Parizi, G. Srivastava, M.-Y. Chen, Cryptocurrency malware hunting: A deep recurrent neural network approach. Appl. Soft Comput. J. Elsevier **96**, 106630 (2020)

25. M. Aledhari, R. Razzak, R.M. Parizi, F. Saeed, Federated learning: A survey on enabling technologies, protocols, and applications. IEEE Access **8**, 140699–140725 (2020). https://doi.org/10.1109/ACCESS.2020.3013541

26. A. Yazdinejad, R.M. Parizi, A. Dehghantanha, H. Karimipour, G. Srivastava, M. Aledhari, Enabling drones in the internet of things with decentralized blockchain-based security. IEEE Internet Things J., 1 (2020). https://doi.org/10.1109/jiot.2020.3015382

27. V. Mothukuri, R.M. Parizi, S. Pouriyeh, Y. Huang, A. Dehghantanha, G. Srivastava, A survey on security and privacy of federated learning. Futur. Gener. Comput. Syst. **115**, 619–640 (2020)

28. A. Yazdinejad, A. Bohlooli, K. Jamshidi, Performance improvement and hardware implementation of Open Flow switch using FPGA, in *IEEE 5th Conference on Knowledge Based Engineering and Innovation, KBEI 2019*, (2019), pp. 515–520

29. A. Singh, K. Click, R.M. Parizi, Q. Zhang, A. Dehghantanha, K.-K.R. Choo, Sidechain technologies in blockchain networks: An examination and state-of-the-art review. J. Netw. Comput. Appl. **149**, 102471 (2020). https://doi.org/10.1016/j.jnca.2019.102471

30. A. Yazdinejad, S. Kavei, S.R. Karizno, Increasing the performance of reactive routing protocol using the load balancing and congestion control mechanism in MANET. Comput. Knowl. Eng. **2**(1), 33–42 (2019). https://doi.org/10.22067/cke

31. D. Połap, G. Srivastava, A. Jolfaei, R.M. Parizi, Blockchain technology and neural networks for the internet of medical things, in *IEEE INFOCOM 2020 – IEEE Conference on Computer Communications Workshops (INFOCOM WKSHPS)*, (2020), pp. 508–513. https://doi.org/10.1109/INFOCOMWKSHPS50562.2020.9162735

32. A. Yazdinejad, R.M. Parizi, A. Dehghantanha, Q. Zhang, K.-K.R. Choo, An energy-efficient SDN controller architecture for IoT networks with blockchain-based security. IEEE Trans. Serv. Comput. (2020). https://doi.org/10.1109/TSC.2020.2966970

33. A. Yazdinejad, G. Srivastava, R.M. Parizi, A. Dehghantanha, K.-K.R. Choo, M. Aledhari, Decentralized authentication of distributed patients in hospital networks using blockchain. IEEE J. Biomed. Heal. Inform. **24**(8), 2146–2156 (2020)

34. A. Yazdinejad, A. Bohlooli, K. Jamshidi, Efficient design and hardware implementation of the OpenFlow v1.3 Switch on the Virtex-6 FPGA ML605. J. Supercomput. **74**(3) (2018). https://doi.org/10.1007/s11227-017-2175-7

35. M. Conti, A. Dehghantanha, K. Franke, S. Watson, Internet of Things security and forensics: Challenges and opportunities. Futur. Gener. Comput. Syst. **78**, 544–546 (2018). https://doi.org/10.1016/j.future.2017.07.060

36. S. Watson, A. Dehghantanha, Digital forensics: The missing piece of the Internet of Things promise. Comput. Fraud Secur. **2016**(6), 5–8 (2016). https://doi.org/10.1016/s1361-3723(15)30045-2

37. H. Darabian et al., Detecting cryptomining malware: A deep learning approach for static and dynamic analysis. J. Grid Comput., 1–11 (2020)

38. S. Homayoun et al., DRTHIS: Deep ransomware threat hunting and intelligence system at the fog layer. Futur. Gener. Comput. Syst. **90**, 94–104 (2019). https://doi.org/10.1016/j.future.2018.07.045

39. E.M. Dovom, A. Azmoodeh, A. Dehghantanha, D.E. Newton, R.M. Parizi, H. Karimipour, Fuzzy pattern tree for edge malware detection and categorization in IoT. J. Syst. Archit. **97**, 1–7 (2019)

40. M. Zolotukhin, T. Hämäläinen, Detection of zero-day malware based on the analysis of opcode sequences, in *2014 IEEE 11th Consumer Communications and Networking Conference (CCNC)*, (2014), pp. 386–391

41. H. Darabian, A. Dehghantanha, S. Hashemi, S. Homayoun, K.R. Choo, An opcode-based technique for polymorphic Internet of Things malware detection. Concurr. Comput. Pract. Exp. **32**(6), e5173 (2020)

42. S.M. Tahsien, H. Karimipour, P. Spachos, Machine learning based solutions for security of Internet of Things (IoT): A survey. J. Netw. Comput. Appl., 102630 (2020)

43. A.N. Jahromi et al., An improved two-hidden-layer extreme learning machine for malware hunting. Comput. Secur. **89**, 101655 (2020)

44. J. Su, V.D. Vasconcellos, S. Prasad, S. Daniele, Y. Feng, K. Sakurai, Lightweight classification of IoT malware based on image recognition, in *2018 IEEE 42Nd Annual Computer Software and Applications Conference (COMPSAC)*, vol. 2, (2018), pp. 664–669

45. Y. Ding, X. Xia, S. Chen, Y. Li, A malware detection method based on family behavior graph. Comput. Secur. **73**, 73–86 (2018)

46. S.D. Nikolopoulos, I. Polenakis, A graph-based model for malware detection and classification using system-call groups. J. Comput. Virol. Hacking Tech. **13**(1), 29–46 (2017)

47. M. Farrokhmanesh, A. Hamzeh, Music classification as a new approach for malware detection. J. Comput. Virol. Hacking Tech. **15**(2), 77–96 (2019)
48. G.G. Sundarkumar, V. Ravi, I. Nwogu, V. Govindaraju, Malware detection via API calls, topic models and machine learning, in *2015 IEEE International Conference on Automation Science and Engineering (CASE)*, (2015), pp. 1212–1217
49. C.-I. Fan, H.-W. Hsiao, C.-H. Chou, Y.-F. Tseng, Malware detection systems based on API log data mining, in *2015 IEEE 39th Annual Computer Software and Applications Conference*, vol. 3, (2015), pp. 255–260
50. S. Vemparala, F. Di Troia, V.A. Corrado, T.H. Austin, M. Stamo, Malware detection using dynamic birthmarks, in *Proceedings of the 2016 ACM on International Workshop on Security and Privacy Analytics*, (2016), pp. 41–46
51. S. Huda, J. Abawajy, M. Alazab, M. Abdollalihian, R. Islam, J. Yearwood, Hybrids of support vector machine wrapper and filter based framework for malware detection. Futur. Gener. Comput. Syst. **55**, 376–390 (2016, Feburary). https://doi.org/10.1016/j.future.2014.06.001
52. H. Darabian et al., A multiview learning method for malware threat hunting: Windows, IoT and android as case studies. World Wide Web **23**(2), 1241–1260 (2020)
53. Y. Ding, W. Dai, S. Yan, Y. Zhang, Control flow-based opcode behavior analysis for malware detection. Comput. Secur. **44**, 65–74 (2014)
54. H. Hashemi, A. Azmoodeh, A. Hamzeh, S. Hashemi, Graph embedding as a new approach for unknown malware detection. J. Comput. Virol. Hacking Tech. **13**(3), 153–166 (2017)
55. H. HaddadPajouh, A. Dehghantanha, R. Khayami, K.-K.R. Choo, A deep recurrent neural network based approach for Internet of Things malware threat hunting. Futur. Gener. Comput. Syst. **85**, 88–96 (2018). https://doi.org/10.1016/j.future.2018.03.007
56. A. Azmoodeh, A. Dehghantanha, K.-K.R. Choo, Robust malware detection for internet of (battlefield) things devices using deep eigenspace learning. IEEE Trans. Sustain. Comput. **4**(1), 88–95 (2018)
57. P. Singh, P.A. Meshram, Survey of density based clustering algorithms and its variants, in *2017 International Conference on Inventive Computing and Informatics (ICICI)*, (2017), pp. 920–926
58. Y. Ye, T. Li, D. Adjeroh, S.S. Iyengar, A survey on malware detection using data mining techniques. ACM Comput. Surv. **50**(3), 1–40 (2017)
59. C.A. Visaggio, P. Swathi, F. Di Troia, T.H. Austin, S. Mark, Clustering for malware classification. J. Comput. Virol. Hacking Tech. **1**, 95–107 (2017)
60. U. Narra, F.D. Troia, V.A. Corrado, T.H. Austin, M. Stamp, Clustering versus SVM for malware detection. J. Comput. Virol. Hacking Tech. **12**(4), 213–224 (2016). https://doi.org/10.1007/s11416-015-0253-z
61. Y. Li et al., Experimental study of fuzzy hashing in malware clustering analysis, in *8th Workshop on Cyber Security Experimentation and Test, CSET 2015*, vol. 5, (2015), p. 52
62. S. Har-Peled, S. Mahabadi, Near neighbor: Who is the fairest of them all? in *Advances in Neural Information Processing Systems*, (2019), pp. 13176–13187
63. S.S. Abraham, S.S. Sundaram, Fairness in clustering with multiple sensitive attributes. arXiv Prepr. arXiv1910.05113 (2019)
64. S. Ahmadian, A. Epasto, R. Kumar, M. Mahdian, Clustering without over-representation, in *Proceedings of the 25th ACM SIGKDD International Conference on Knowledge Discovery & Data Mining*, (2019), pp. 267–275
65. I.O. Bercea et al., On the cost of essentially fair clusterings. arXiv Prepr. arXiv1811.10319 (2018)
66. L.E. Celis, L. Huang, V. Keswani, N.K. Vishnoi, Classification with fairness constraints: A meta-algorithm with provable guarantees, in *Proceedings of the Conference on Fairness, Accountability, and Transparency*, (2019), pp. 319–328
67. X. Chen, B. Fain, C. Lyu, K. Munagala, Proportionally fair clustering. arXiv Prepr. arXiv1905.03674 (2019)
68. A.K. Menon, R.C. Williamson, The cost of fairness in binary classification, in *Conference on Fairness, Accountability and Transparency*, (2018), pp. 107–118

Printed in the United States
by Baker & Taylor Publisher Services